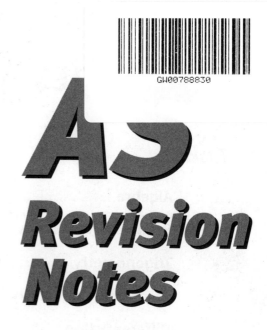

Mathematics

Author

Peter Sherran

Contents

Pure 1

Algebra 5

Sequences and series 16

Trigonometry 18

Co-ordinate geometry 21

Differentiation 24

Integration 27

Progress check 29

Pure 2

Algebra and functions 30

Trigonometry 38

Differentiation 42

Integration 44

Numerical methods 46

Progress check 49

Mechanics 1

Vectors 50

Kinematics 53

Statics 57

Dynamics 62

Progress check 66

Statistics 1

Representing data 67

Probability 69

Correlation and regression 71

Discrete random variables 73

The Normal distribution 76

Progress check 77

Decision mathematics 1

Algorithms 78

Graphs and networks 81

Critical path analysis 87

Linear programming 89

Matchings 92

Progress check 93

Progress check answers 94

Index 96

Pure 1

Algebra

Indices

You need to know all of the basic rules and be able to apply them.

$a^m \times a^n = a^{m+n}$ $\qquad$ $a^m \div a^n = a^{m-n}$ $\qquad$ $(a^m)^n = a^{mn}$ $\qquad$ $a^{\frac{1}{n}} = \sqrt[n]{a}$

$a^{\frac{m}{n}} = (a^m)^{\frac{1}{n}} = (a^{\frac{1}{n}})^m$ $\qquad$ $a^{-n} = \dfrac{1}{a^n}$ $\qquad$ $(ab)^n = a^n b^n$ $\qquad$ $\left(\dfrac{a}{b}\right)^n = \dfrac{a^n}{b^n}$

Some important special cases are: $\quad a^1 = a \quad a^{\frac{1}{2}} = \sqrt{a}$

$\qquad\qquad\qquad\qquad\qquad$ and $\quad a^0 = 1 \quad a^{-1} = \dfrac{1}{a}$ provided $a \neq 0$.

For example $9^{\frac{1}{2}} = \sqrt{9} = 3$ and $16^{\frac{3}{2}} = (16^{\frac{1}{2}})^3 = 4^3 = 64$.

Surds

A **surd** is the square root of a whole number that has an **irrational** value.

Some examples are $\sqrt{2}$, $\sqrt{3}$ and $\sqrt{10}$.
You can often simplify a surd using the fact that $\sqrt{ab} = \sqrt{a} \times \sqrt{b}$.

For example $\sqrt{12} = \sqrt{4 \times 3} = \sqrt{4} \times \sqrt{3} = 2\sqrt{3}$. Surds may be used to express some results in exact form such as $\sin 60° = \dfrac{\sqrt{3}}{2}$.

Functions

A **function** may be thought of as a rule which takes each member x of a set and assigns, or **maps**, it to some value y known as its **image**.

$$x \longrightarrow \boxed{\text{Function}} \longrightarrow y$$

A letter such as f, g or h is often used to stand for a function. The function which squares a number and adds on 5, for example, can be written as $f(x) = x^2 + 5$. The same notation may also be used to show how a function affects particular values.

For this function, $f(4) = 4^2 + 5 = 21$, $f(-10) = (-10)^2 + 5 = 105$ and so on.

Examiner's Tip

An alternative notation for the same function is $f: x \mapsto x^2 + 5$.

Algebra

The set of values on which the function acts is called the **domain** and the corresponding set of image values is called the **range**.

The function $f(x) = x + 2$ with domain $\{3, 6, 10\}$ is shown below.

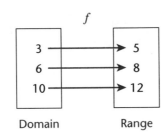

The domain of a function may be an infinite set such as $\mathbb{R}$ (the real numbers) but, in some cases, particular values must be omitted for the function to be valid.

For example, the function $f(x) = \dfrac{x+2}{x-3}$ can not have 3 in its domain since division by zero is undefined.

Under a function, every member of the domain must have only one image. But, it is possible for different values in the domain to have the same image.

If for each element y in the range, there is *a unique* value of x such that $f(x) = y$ then f is a **one–one** function. If for any element y in the range, there is more than one value of x satisfying $f(x) = y$ then f is **many–one**.

Examples

The function $f(x) = 2x$ with domain $\mathbb{R}$ is one–one.

The function $f(x) = x^2$ with domain $\mathbb{R}$ is many–one.

The function $f(x) = x^2$ with domain $x > 0$ is one–one.

The function $f(x) = \sin x$ with domain $0° \leqslant x \leqslant 360°$ is many–one.

The function $f(x) = \sin x$ with domain $-90° \leqslant x \leqslant 90°$ is one–one.

Quadratics

In algebra, any expression of the form $ax^2 + bx + c$ where $a \neq 0$ is called a **quadratic**.

You need to know these basic results about quadratics.

$$(x+a)^2 = x^2 + 2ax + a^2 \qquad (x-a)^2 = x^2 - 2ax + a^2 \qquad (x+a)(x-a) = x^2 - a^2$$

Some examples are:

$$(x+3)^2 = x^2 + 6x + 9 \qquad (x-5)^2 = x^2 - 10x + 25$$
$$(x+7)(x-7) = x^2 - 49 \qquad (x+\sqrt{3})(x-\sqrt{3}) = x^2 - 3.$$

To solve problems in algebra you need to develop your skills so that you can recognise how to apply the basic results.

For example, the fraction $\dfrac{1}{2-\sqrt{3}}$ may be simplified by **rationalising the denominator**.

You can do this by multiplying the numerator and denominator by $2+\sqrt{3}$ to make an equivalent fraction in which the denominator is rational.

$$\frac{1}{2-\sqrt{3}} = \frac{1}{2-\sqrt{3}} \times \frac{2+\sqrt{3}}{2+\sqrt{3}}$$

$$= \frac{2+\sqrt{3}}{4-3} \qquad \text{(The denominator is now rational)}$$

$$= 2+\sqrt{3}.$$

Quadratic equations

Equations of the form $ax^2 + bx + c = 0$ (where $a \neq 0$) are quadratic equations.

Some quadratic equations can be solved by **factorising**.

Example Solve $x^2 - 3x - 10 = 0$.
$(x-5)(x+2) = 0$ This means that either $x - 5 = 0$
                                              or $x + 2 = 0$.
so $x = 5$ or $x = -2$.

If the quadratic will not factorise then you can try **completing the square**.

Example Solve $x^2 - 6x + 1 = 0$.
$(x-3)^2 - 8 = 0$ (Now x only appears once in the equation)
$(x-3)^2 = 8$
$x - 3 = \pm\sqrt{8}$
$x = 3 \pm 2\sqrt{2}$.   (The solutions are $3 + 2\sqrt{2}$ and $3 - 2\sqrt{2}$)

Examiner's Tip

The method shown for completing the square can be adapted for the general form of a quadratic. This gives the usual formula for the solution of a quadratic equation.

Algebra

The quadratic formula

$ax^2 + bx + c = 0$

$4a^2x^2 + 4abx + 4ac = 0$ Note: $4a^2x^2 + 4abx + b^2 = (2ax + b)^2$.

$4a^2x^2 + 4abx + b^2 + 4ac - b^2 = 0$

$(2ax + b)^2 + 4ac - b^2 = 0$ Now x only appears once in the equation.

$(2ax + b)^2 = b^2 - 4ac$

$2ax + b = \pm\sqrt{b^2 - 4ac}$

$x = \dfrac{-b \pm \sqrt{b^2 - 4ac}}{2a}$ This is the **quadratic formula**.

Example Solve $3x^2 - 4x - 9 = 0$ giving your answers to two decimal places.

Comparing this equation with the general form gives $a = 3$, $b = -4$ and $c = -9$. Substitute this information into the formula:

$$x = \frac{4 \pm \sqrt{(-4)^2 - 4 \times 3 \times (-9)}}{6} = \frac{4 \pm \sqrt{124}}{6}$$

$x = 2.52$ or -1.19 (two decimal places).

In the formula, the value of $b^2 - 4ac$ is called the **discriminant**. This value can be used to give information about the solutions without having to solve the equation.

$b^2 - 4ac > 0$ *two* real solutions (the solutions are often called **roots**)
$b^2 - 4ac = 0$ *one* real solution (often thought of as one **repeated root**)
$b^2 - 4ac < 0$ *no* real solutions.

The solutions of a quadratic equation correspond to where its graph crosses the x-axis. There are three possible situations depending on the value of the discriminant.

These diagrams correspond to the situation when $a > 0$.

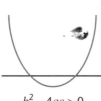
$b^2 - 4ac > 0$

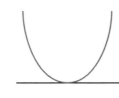
$b^2 - 4ac = 0$

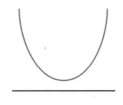
$b^2 - 4ac < 0$

Example The equation $5x^2 + 3x + p = 0$ has a repeated root. Find the value of p.

In this case, $a = 5$, $b = 3$ and $c = p$.
For a repeated root $b^2 - 4ac = 0$ so $9 - 20p = 0$, giving $p = 0.45$.

Quadratic graph

The graph of $y = ax^2 + bx + c$ is a parabola.

$a > 0 \qquad\qquad a < 0$

The methods used for solving quadratic equations can also be used to give information about the graphs.

Example Sketch the graph of $y = x^2 - x - 6$.
Find the co-ordinates of the lowest point on the curve.

The curve will cross the x-axis when $y = 0$. You can find these points by solving the equation $x^2 - x - 6 = 0$.

$x^2 - x - 6 = 0 \Rightarrow (x + 2)(x - 3) = 0$

$\Rightarrow x = -2 \quad \text{or} \quad x = 3$.

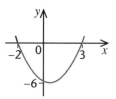

The curve is symmetrical so the lowest point occurs mid-way between -2 and 3 and this is given by $(-2 + 3) \div 2 = 0.5$.

When $x = 0.5$, $y = 0.5^2 - 0.5 - 6 = -6.25$.

The lowest point on the curve is $(0.5, -6.25)$.

Examiner's Tip

Completing the square gives information about the **vertex** of the curve even if the equation will not factorise.

Mathematics Revision Notes

Algebra

Example Find the co-ordinates of the vertex of the curve $y = x^2 + 2x + 3$.

You need to recognise that $x^2 + 2x + 1 = (x + 1)^2$,

then, completing the square, $x^2 + 2x + 3 = x^2 + 2x + 1 + 2 = (x + 1)^2 + 2$.

The equation of the curve can now be written as $y = (x + 1)^2 + 2$.

$(x + 1)^2$ cannot be negative so it's minimum value is zero, when $x = -1$.

This means that the minimum value of y is 2 and this occurs when $x = -1$.

The vertex of the curve is at $(-1, 2)$.

More graphs

The graph of a cubic equation $y = ax^3 + bx^2 + cx + d$ can take a number of forms.

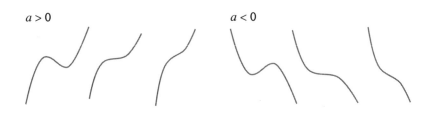

The cubic curve has rotational symmetry of order 2 about the point where $x = -\dfrac{b}{3a}$.

A cubic equation that can be factorised as $y = (x - p)(x - q)(x - r)$ will cross the x-axis at p, q and r. If any two of p, q and r are the same then the x-axis will be a tangent to the curve at that point.

For example, the graph of $y = (x + 2)(x - 3)^2$ looks like this:

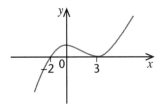

This shows a **repeated root** at $x = 3$.

Examiner's Tip

The graph of $y = x^n$ where n is an integer has:
rotational symmetry about the **origin** when n is **odd**
reflective symmetry in the y-axis when n is **even**.

The graphs of $y = x^3$, $y = x^5$, $y = x^7$ ... look something like this:

The graphs of $y = x^2$, $y = x^4$, $y = x^6$ ... look something like this:

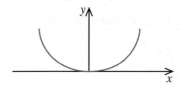

For higher powers of x the graphs are flatter between -1 and 1 and steeper elsewhere.

The graphs of $y = x^{-1}$, $y = x^{-3}$, $y = x^{-5}$ ... look something like this:

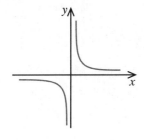

The graphs of $y = x^{-2}$, $y = x^{-4}$, $y = x^{-6}$ ... look something like this:

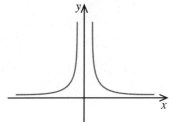

As the powers of x become more negative, the graphs become steeper between -1 and 1 and flatter elsewhere.

The modulus function

The **modulus function** $y = |ax + b|$ represents the positive value of the expression inside the brackets. The graph of $y = |ax + b|$ is identical to the graph of $y = ax + b$ when $ax + b \geq 0$ but is given by its reflection in the x-axis when $ax + b < 0$.

For example, compare the graphs of $y = x - 5$ and $y = |x - 5|$.

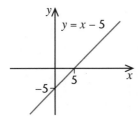
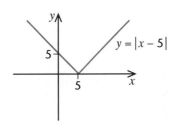

Examiner's Tip

To sketch the graph of $y = |ax + b|$, start with the sketch of $y = ax + b$ and reflect any parts that are below the x-axis.

Algebra

Transforming graphs

The graph of some new function can often be obtained from the graph of a known function by applying a transformation. A summary of the standard transformations is given in the table.

Known function	New function	Transformation
$y = f(x)$	$y = f(x) + a$	Translation through a units parallel to y-axis.
	$y = f(x - a)$	Translation through a units parallel to x-axis.
	$y = af(x)$	One-way stretch with scale factor a parallel to the y-axis.
	$y = f(ax)$	One-way stretch with scale factor $\frac{1}{a}$ parallel to the x-axis.

Example The diagram shows the graph of the function $y = f(x)$ where

$$f(x) = 1 - |x - 2| \quad \text{for} \quad 1 \leqslant x \leqslant 3.$$

Use the same axes to show:
(a) $y = f(x) + 1$
(b) $y = f(x + 1)$
(c) $y = 2f(x)$
(d) $y = f(2x)$

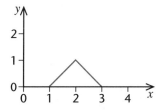

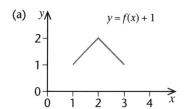

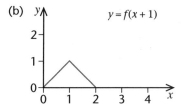

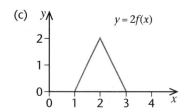

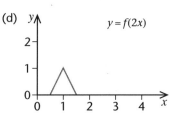

Examiner's Tip

You may need to apply a combination of transformations in some cases.

Simultaneous equations

The substitution method of solving pairs of linear simultaneous equations can also be applied when one of the equations is a quadratic.

Example Find the co-ordinates of the points where the line $y = 2x + 3$ intersects the circle $x^2 + y^2 = 26$.

The co-ordinates at the points of intersection satisfy both equations simultaneously.

Substitute for y in the quadratic equation: $\quad x^2 + (2x + 3)^2 = 26.$

Now remove the brackets: $\quad x^2 + 4x^2 + 12x + 9 = 26.$

Simplify and write in the form $ax^2 + bx + c = 0$: $\quad 5x^2 + 12x - 17 = 0.$

Use the formula to find x:
$$x = \frac{-12 \pm \sqrt{144 + 340}}{10}.$$
$x = 1$ or $x = -3.4$.

Substitute the x values in the linear equation to find the corresponding y values: $\quad y = 5$ or $y = -3.8$.

The co-ordinates at the points of intersection are $(1,5)$ and $(-3.4,-3.8)$.

Identities and equations

Identities and equations are closely related but there is an important difference. To make the point consider these two statements.

(1) $x^2 = 3x - 2$
(2) $(x + 5)(x - 5) = x^2 - 25.$

The first statement is an **equation** because it is only true for particular values of x. The second statement is an **identity** because it is true for *all* values of x.

Depending on the context, the symbol $\equiv$ may be used for an identity in place of $=$ to emphasise its significance.

Example Find the value of A, B and C such that:

$2x^2 - 9x - 15 \equiv Ax(x + 3) + B(x + 3)^2 + C(x^2 + 1).$

Since this works for all values of x, you can substitute values of your choice to make simultaneous equations involving A, B and C.

Substituting $x = -3$ gives: $\quad 30 = 10C \Rightarrow C = 3.$

Substituting $x = 0$ gives: $\quad -15 = 9B + C \quad$ but $C = 3 \quad$ so $-15 = 9B + 3 \Rightarrow B = -2.$

Equating the coefficients of x^2 gives: $\quad 2 = A + B + C, \quad$ so $2 = A - 2 + 3 \Rightarrow A = 1.$

Examiner's Tip

Sometimes, you run out of convenient values to use. Another approach that works for identities is to match the coefficients of particular terms on each side.

Algebra

Polynomials and the factor theorem

A quadratic is one example of a **polynomial**. In general, a polynomial takes the form:

$$a_n x^n + a_{n-1} x^{n-1} + a_{n-2} x^{n-2} + \ldots + a_0,$$

where $a_n, a_{n-1}, \ldots a_0$ are constants and n is a positive whole number.

For example, $x^4 - 2x^3$ is a polynomial. In this case $a_4 = 1$, $a_3 = -2$ and a_2, a_1, a_0 are all zero. The **degree** of a polynomial is the highest power of x that it includes, so the degree of $x^4 - 2x^3$ is 4. A quadratic is a polynomial of degree 2, a cubic is a polynomial of degree 3 and so on.

One important result that holds for a polynomial of any order is the **factor theorem**. This states that if $f(x)$ is a polynomial and $f(a) = 0$ then $(x - a)$ is a factor of $f(x)$.

Example A polynomial is given by $f(x) = 2x^3 + 13x^2 + 13x - 10$.
(a) Find the value of $f(2)$ and $f(-2)$.
(b) State one of the factors of $f(x)$.
(c) Factorise $f(x)$ completely.

(a) $f(2) = 2(2^3) + 13(2^2) + 13(2) - 10 = 84$

$f(-2) = 2(-2)^3 + 13(-2)^2 + 13(-2) - 10 = 0.$

(b) Since $f(-2) = 0$, $(x + 2)$ is a factor of $f(x)$ by the factor theorem.

(c) $2x^3 + 13x^2 + 13x - 10 \equiv (x + 2)(ax^2 + bx + c).$

This is an **identity** and so the values of a, b and c may be found by comparing coefficients.

Comparing the x^3 terms gives: $a = 2.$
Comparing the x^2 terms gives: $13 = 2a + b \Rightarrow b = 9.$
Comparing the constant terms gives: $c = -5.$
It follows that: $f(x) = (x + 2)(2x^2 + 9x - 5).$

Factorising the quadratic part in the usual way gives $f(x) = (x + 2)(2x - 1)(x + 5).$

Inequalities

Linear inequalities can be solved by rearrangement in much the same way as linear equations. However, care must be taken to reverse the direction of the inequality when multiplying or dividing by a negative.

Example Solve the inequality $8 - 3x > 23$.

Subtract 8 from both sides: $\quad -3x > 15$
Divide both sides by -3: $\quad\quad x < -5$.

An inequality which has x on both sides is treated like the corresponding equation.

Example Solve the inequality $5x - 3 > 3x - 10$.

Subtract $3x$ from both sides: $\quad 2x - 3 > -10$
Add 3 to both sides: $\quad\quad\quad\quad\quad 2x > -7$
Divide both sides by 2: $\quad\quad\quad\quad x > -3.5$.

Quadratic inequalities are solved in a similar way to quadratic equations but a sketch graph is often helpful at the final stage.

Example Solve the inequality $x^2 - 3x + 2 < 0$.

Factorise the quadratic expression: $(x-1)(x-2) < 0$.

Sketch the graph of $y = (x-1)(x-2)$:

The graph shows that $(x-1)(x-2) < 0$ between 1 and 2.

It follows that $x^2 - 3x + 2 < 0$ when $1 < x < 2$.

Examiner's Tip

If the quadratic expression cannot be factorised then the formula may be used to find the points of intersection of the graph with the x-axis.

Sequences and series

Sequences

A list of numbers in a particular order, that follow some rule for finding later values, is called a **sequence**. Each number in a sequence is called a **term**, and terms are often denoted by $u_1, u_2, u_3, ..., u_n, ...$.

One way to define a sequence is to give a formula for the nth term such as $u_n = n^2$. Substituting the values $n = 1, 2, 3, 4, ...$ produces the sequence 1, 4, 9, 16, ... and the value of any particular term can be calculated by substituting its position number into the formula. For example, in this case, the 50th term is $u_{50} = 50^2 = 2500$.

Another way to define a sequence is to give a starting value together with a rule that shows the connection between successive terms. This is sometimes called a **recursive definition**. For example $u_1 = 5$ and $u_{n+1} = 2u_n$ defines the sequence 5, 10, 20, 40, ... The rule $u_{n+1} = 2u_n$ is an example of a **recurrence relation**.

Two special sequences are the **arithmetic progression** (A.P.) and the **geometric progression** (G.P.). In an A.P. successive terms have a common difference, e.g. 1, 4, 7, 10, ... The first term is denoted by a and the common difference is d. With this notation, the definition of an A.P. may now be given as $u_1 = a$, $u_{n+1} = u_n + d$.

The terms of an A.P. take the form $a, a + d, a + 2d, a + 3d, ...$ and the nth term is given by $u_n = a + (n - 1)d$.

In a G.P. successive terms are connected by a **common ratio**, e.g. 3, 6, 12, 24, ... The definition of a G.P. may be given as $u_1 = a$, $u_{n+1} = ru_n$ where r is the common ratio. The terms of a G.P. take the form $a, ar, ar^2, ar^3, ...$ and the nth term is given by $u_n = ar^{n-1}$.

Series

A series is formed by adding together the terms of a sequence. The use of sigma notation can greatly simplify the way that series are written. For example, the series $1^2 + 2^2 + 3^2 + ... + n^2$ may be written as $\sum_{i=1}^{n} i^2$. The sum of the first n terms of a series is often denoted by S_n and so $S_n = u_1 + u_2 + u_3 + ... + u_n = \sum_{i=1}^{n} u_i$.

The **sum of an A.P.** is given by $S_n = a + (a + d) + (a + 2d) + ... + (a + (n - 1)d)$.
This may also be written as $S_n = l + (l - d) + (l - 2d) + ... + (l - (n - 1)d)$, where l is the last term.

Adding the two versions gives $2S_n = (a + l) + (a + l) + (a + l) + ... (a + l)$
$= n(a + l)$.

This gives the general result: $S_n = \dfrac{n}{2}(a + l)$.

Substituting $l = a + (n-1)d$ gives the alternative form of the result:
$$S_n = \frac{n}{2}(2a + (n-1)d).$$

A special case is the sum of the first n natural numbers:
$$1 + 2 + 3 + \ldots + n = \frac{n}{2}(n+1).$$

For example, the sum of the first 100 natural numbers is $\frac{100}{2}(100 + 1) = 5050$.

When finding the sum of an A.P. you need to select the most appropriate version of the formula to suit the information.

Example Find the sum of the first 50 terms of the series $15 + 18 + 21 + 24 + \ldots$.

In this series, $a = 15$, $d = 3$ and $n = 50$.

Using $S_n = \frac{n}{2}(2a + (n-1)d)$ gives $S_{50} = \frac{50}{2}(2 \times 15 + 49 \times 3) = 4425$.

The sum of a G.P. is given by $S_n = a + ar + ar^2 + ar^3 + \ldots + ar^{n-1}$.
Multiplying throughout by r gives $rS_n = ar + ar^2 + ar^3 + \ldots + ar^{n-1} + ar^n$.
Subtracting gives: $S_n - rS_n = a - ar^n$
$$\Rightarrow S_n(1 - r) = a(1 - r^n)$$

So the sum of the first n terms of a G.P. is $S_n = \dfrac{a(1 - r^n)}{1 - r}$.

If $r > 1$ then it is more convenient to use the result in the form $S_n = \dfrac{a(r^n - 1)}{r - 1}$.

Example Find the sum of the first 20 terms of the series $8 + 12 + 18 + 27 + \ldots$ to the nearest whole number.

In this series, $a = 8$, $r = 1.5$ and $n = 20$.

This gives $S_{20} = \dfrac{8(1.5^{20} - 1)}{1.5 - 1} = 53188.107\ldots = 53188$ to the nearest whole number.

Provided that $|r| < 1$ the sum of a G.P. converges to $\dfrac{a}{1-r}$ as n tends to infinity.

For example, $1 + \dfrac{1}{2} + \dfrac{1}{4} + \dfrac{1}{8} + \ldots = \dfrac{1}{1 - \frac{1}{2}} = 2$.

Examiner's Tip

You need to know how to establish the general results. You may be asked to do this in the exam.

Trigonometry

The sine, cosine and tangent of 30°, 45°, and 60° may be expressed exactly as shown below. The results are based on Pythagoras' theorem.

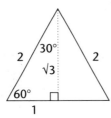

$\sin 30° = \dfrac{1}{2}$ $\sin 60° = \dfrac{\sqrt{3}}{2}$ $\sin 45° = \dfrac{1}{\sqrt{2}}$

$\cos 30° = \dfrac{\sqrt{3}}{2}$ $\cos 60° = \dfrac{1}{2}$ $\cos 45° = \dfrac{1}{\sqrt{2}}$

$\tan 30° = \dfrac{1}{\sqrt{3}}$ $\tan 60° = \sqrt{3}$ $\tan 45° = 1$.

$y = \sin x$

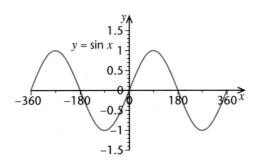

sin x is defined for any angle and always has a value between −1 and 1. It is a **periodic function** with period 360°.

The graph has **rotational symmetry** of order 2 about the origin and about every point where it crosses the x-axis.

It has **line symmetry** about every vertical line passing through a vertex.

$y = \cos x$

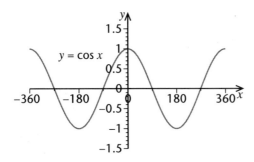

$\cos x \equiv \sin(x + 90°)$ so the graph of $y = \cos x$ can be obtained by translating the sine graph 90° to the left.

It follows that cos x is also a periodic function with period 360° and has the corresponding symmetry properties.

$y = \tan x$

$\tan x \equiv \dfrac{\sin x}{\cos x}$.

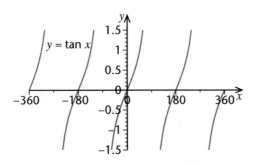

tan x is undefined whenever cos $x = 0$ and approaches ±∞ near these values. It is a periodic function with period 180°.

The graph has rotational symmetry of order 2 about 0°, ±90°, ±180°, ±270°, … .

The graphs of more complex trigonometric functions can often be produced by applying transformations to one of the basic graphs.

Example Sketch the graph of $y = \sin(2x + 30) + 1$ for $0° \leq x \leq 360°$.

It is helpful to think about building the transformations in stages.

	Transformations		
Basic function	Apply a one-way stretch with scale factor 2 parallel to the x-axis.	Now translate the curve 30° to the left.	Finally translate the curve 1 unit up.
$y = \sin x$	$y = \sin(2x)$	$y = \sin(2x + 30)$	$y = \sin(2x + 30) + 1$

The graph of $y = \sin(2x + 30°) + 1$ may now be produced by applying these transformations in order starting from the graph of $y = \sin x$.

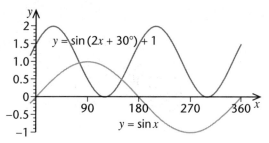

The identities $\tan x \equiv \dfrac{\sin x}{\cos x}$ and $\sin^2 x + \cos^2 x \equiv 1$ are very important in terms of simplifying expressions and solving equations.

Example   (a) Simplify $1 - (\sin x - \cos x)^2$

(b) Solve $2 \cos x = 5 \sin x$ for $-360° \leq x \leq 360°$.

(a) $1 - (\sin x - \cos x)^2 \equiv 1 - \sin^2 x + 2 \sin x \cos x - \cos^2 x$
$\equiv 1 - (\sin^2 x + \cos^2 x) + 2 \sin x \cos x$
$\equiv 2 \sin x \cos x.$

(b) $2 \cos x = 5 \sin x \Rightarrow 2 = \dfrac{5 \sin x}{\cos x}$
$\Rightarrow 2 = 5 \tan x$
$\Rightarrow \tan x = \tfrac{2}{5}.$

One solution is $x = \tan^{-1}(\tfrac{2}{5}) = 21.8°$ to 1 d.p.

Examiner's Tip

To find the other solutions in the interval $-360° \leq x \leq 360°$ you can look at the sketch of $y = \tan x$.

Trigonometry

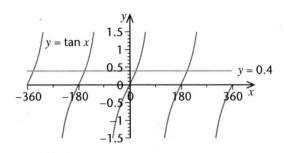

The graph shows that there are four solutions in the required interval. These values are 180° apart.

The solutions are:
201.8°, 21.8°, −158.2°, −338.2°.

You may be required to solve trigonometric equations in a variety of forms.

Example Solve $\sin(2x - 30°) = 0.6$ for $0° \leq x \leq 360°$.

One value of $2x - 30°$ is $\sin^{-1}(0.6) = 36.86...°$.
You need to adjust the interval to find where the values of $2x - 30°$ must lie.

$$0° \leq x \leq 360° \Rightarrow 0° \leq 2x \leq 720° \Rightarrow -30° \leq 2x - 30° \leq 690°.$$

Now look at the graph of the sine function in this interval.

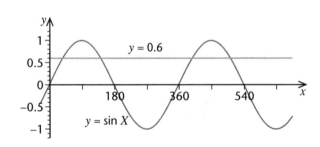

This is the graph of $y = \sin X$ for $-30° \leq X \leq 690°$ where $X = 2x - 30$. In other words, draw the sine curve as normal for the interval and then interpret the variable as $2x - 30$. The graph shows that there are 4 values of $2x - 30°$ in the interval that make the equation work.

Using the symmetry of the curve, the values of $2x - 30°$ are:
36.86...°, 180° − 36.86...°, 360° + 36.86...°, 540° − 36.86...°.

This gives:
$2x - 30° = 36.86...° \Rightarrow x = 33.4°$; $2x - 30° = 143.14...° \Rightarrow x = 86.6°$.
$2x - 30° = 396.86...° \Rightarrow x = 213.4°$; $2x - 30° = 503.14...° \Rightarrow x = 266.6°$.

Radians

Angles can also be measured in **radians** and this makes it much easier to deal with trigonometric functions when using **calculus**.

1 radian ≈ 57.3° this may be written as $1^c \approx 57.3°$. However, the symbol for radians is not normally written when the angle involves π. The following results are useful to remember: $\pi = 180°$, $\frac{\pi}{2} = 90°$, $\frac{\pi}{4} = 45°$, $\frac{\pi}{3} = 60°$, $\frac{\pi}{6} = 30°$.

Examiner's Tip

All of the results that you know in degrees also apply in radians, e.g. $\sin \frac{\pi}{6} = 0.5$.

Co-ordinate geometry

Straight lines

The general equation of a straight line is $ax + by + c = 0$. The equation of any straight line can be written in this form.

For example, the line $x + y = 6$ corresponds to $a = 1$, $b = 1$ and $c = -6$.

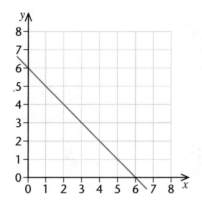

Straight lines, apart from those parallel to the y-axis, can be written in the form $y = mx + c$. This is known as **gradient–intercept** form because the gradient (m) and the y-intercept (c) are clearly identified in the equation. This makes it easy to construct an equation when the gradient and intercept are known.

For example, the line with gradient 4 crossing the y-axis at -5 has equation $y = 4x - 5$.

Straight lines that are **parallel** must have the **same gradient**. The gradient of a vertical line is undefined.

Example Find the equation of the straight line parallel to $y = 3x - 5$ and passing through the point $(4, 2)$.

The line must have gradient 3 and so it can be written in the form $y = 3x + c$. Substituting $x = 4$ and $y = 2$ gives $2 = 3 \times 4 + c \Rightarrow c = -10$.

The required equation is $y = 3x - 10$.

In general the equation of a straight line with gradient m and passing through the point (x_1, y_1) can be written as $y - y_1 = m(x - x_1)$.

Examiner's Tip

Using $y - y_1 = m(x - x_1)$ in the example above gives $y - 2 = 3(x - 4)$. The equation is acceptable in this form but it can be rearranged to give $y = 3x - 10$ as before.

Co-ordinate geometry

Example Find the equation of the straight line passing through the points $(-1,5)$ and $(3,-2)$.

One approach is to use the form $y = mx + c$ to produce a pair of simultaneous equations:

Substituting $x = -1$ and $y = 5$ gives $\qquad 5 = -m + c \qquad (1)$
Substituting $x = 3$ and $y = -2$ gives $\qquad -2 = 3m + c \qquad (2)$

(2) − (1) gives: $\qquad -7 = 4m \Rightarrow m = \dfrac{-7}{4}$.

Substituting for m in (1) gives: $\qquad 5 = \dfrac{7}{4} + c \Rightarrow c = \dfrac{13}{4}$.

The equation of the line is $y = \dfrac{-7}{4}x + \dfrac{13}{4}$. This is the same as $4y + 7x = 13$.

An alternative approach is to find the gradient directly and then use the form $y - y_1 = m(x - x_1)$.

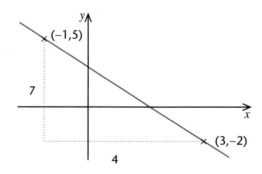

From the diagram, the line has gradient $\dfrac{-7}{4}$.

The equation of the line is $y - 5 = \dfrac{-7}{4}(x + 1)$.

To show that this is the same as the previous result (you don't *need* to do this):

Multiply both sides of the equation by 4: $\qquad 4y - 20 = -7(x + 1)$.
Expand the brackets: $\qquad 4y - 20 = -7x - 7$.
Rearrange to give the previous result: $\qquad 4y + 7x = 13$.

An equation of the form $ax + by + c = 0$ can be rearranged into gradient–intercept form provided that $b \neq 0$. This becomes $y = -\frac{a}{b}x - \frac{c}{b}$ and shows that parallel lines may be produced by keeping a and b fixed and allowing c to change.

For example, the lines $2x - 3y = 5$, $2x - 3y = -2$, $4x - 6y = 7$ are all parallel.

When two straight lines are **perpendicular**, the **product** of their **gradients** is −1.

Example Find the equation of the straight line perpendicular to the line $4x + 3y = 12$ and passing through the point $(2,5)$.

Rearrange the equation into gradient–intercept form: $\quad y = -\tfrac{4}{3}x + 4$.

If the gradient of the required line is m then: $\quad m \times -\tfrac{4}{3} = -1 \implies m = \tfrac{3}{4}$.

Using the form $y - y_1 = m(x - x_1)$ gives: $\quad y - 5 = \tfrac{3}{4}(x - 2)$.

If P has co-ordinates (x_1, y_1) and Q has co-ordinates (x_2, y_2) then the **mid-point** of PQ has co-ordinates $\left(\dfrac{x_1 + x_2}{2}, \dfrac{y_1 + y_2}{2}\right)$.

Example The mid-point of $A(-3,1)$ and $B(5,7)$ is C.

Find the co-ordinates of C.

The co-ordinates of C are given by $\left(\dfrac{-3 + 5}{2}, \dfrac{1 + 7}{2}\right) = (1,4)$.

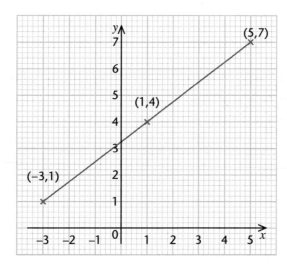

Examiner's Tip

It is often a good idea to draw a diagram if you have time. It may help you interpret a question or verify the theoretical result.

Differentiation

The **gradient** of a curve changes continuously along its length. Its value at any point P is given by the gradient of the tangent to the curve at P. The gradient may be found approximately by drawing.

The exact value of the gradient is found by **differentiation**. This is the **limit** of the gradient of PQ as Q moves towards P.

Gradient of curve at P $\approx \dfrac{\delta y}{\delta x}$.

As Q moves towards P, $\delta x \to 0$ and $\dfrac{\delta y}{\delta x} \to \dfrac{dy}{dx}$.

$\dfrac{dy}{dx}$ is the **gradient function** and represents the **derivative** of y with respect to x.

The graph of $y = kx^n$ has gradient function $\dfrac{dy}{dx} = nkx^{n-1}$.

Example Find the gradient of the curve $y = 3x^2$ at the point P (5,75).

$\dfrac{dy}{dx} = 2 \times 3x^1 = 6x$. At P: $x = 5$ so the gradient is $6 \times 5 = 30$.

Function notation may also be used for derivatives. If $y = f(x)$ then $\dfrac{dy}{dx} = f'(x)$.

The notation is useful for stating some of the rules of differentiation.

$$\text{If } y = f(x) \pm g(x) \text{ then } \dfrac{dy}{dx} = f'(x) \pm g'(x)$$

Example $f(x) = x^3 + 5x + 2$. Find (a) $f'(x)$ (b) $f'(4)$.

(a) $f'(x) = 3x^2 + 5$ (b) $f'(4) = 3 \times 4^2 + 5 = 53$.

You often need to express a function in the right form before you can differentiate it.

Example Differentiate: (a) $f(x) = \sqrt{x}$ (b) $f(x) = (x-5)(x+3)$ (c) $f(x) = \dfrac{x^3+1}{x^2}$.

(a) $\sqrt{x}$ has to be written as a power of x to use the rule $f'(x) = nkx^{n-1}$.

$\sqrt{x} = x^{\frac{1}{2}}$ so $f(x) = x^{\frac{1}{2}}$ and $f'(x) = \tfrac{1}{2}x^{-\frac{1}{2}}$.

(b) The brackets must first be removed and *then* you can differentiate term by term.

$$f(x) = (x-5)(x+3) = x^2 - 2x - 15.$$

Now $f'(x) = 2x - 2$.

(c) Divide $x^3 + 1$ by x^2 first giving $f(x) = x + x^{-2}$.

Now, $f'(x) = 1 - 2x^{-3} = 1 - \dfrac{2}{x^3}$.

> Differentiate each term separately.
> Remember that $x^0 = 1$.
> When you differentiate a constant, the result is zero.

Tangents and normals

You can find the gradient of the tangent to a curve at a point by differentiation. Then you can use the techniques described in the Co-ordinate Geometry section to find the **equation of the tangent** and the **equation of the normal** at the given point.

Example Find the equation of the tangent and the normal to the curve $y = x^3 - 4x$ at the point (2,0).

Differentiate the equation of the curve to give $\dfrac{dy}{dx} = 3x^2 - 4$.

The gradient of the tangent at (2,0) is $3 \times 2^2 - 4 = 8$.
Using $y - y_1 = m(x - x_1)$ gives the equation of the tangent as $y = 8(x - 2)$.
The gradient of the normal is $-\tfrac{1}{8}$. Using $y - y_1 = m(x - x_1)$ again gives the equation of the normal as $y = -\tfrac{1}{8}(x - 2)$. You could rearrange this to give $x + 8y = 2$.

Curve sketching

The gradient function gives information about the behaviour of a curve.

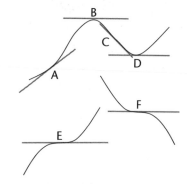

$f'(x) > 0$ at A and the function is increasing.

$f'(x) < 0$ at C and the function is decreasing.

$f'(x) = 0$ at B and D and the function is neither increasing nor decreasing. These points are called **stationary points**. B is at a **local maximum** and D is at a **local minimum**.

E and F show a different type of stationary point called a **stationary point of inflexion**.

Examiner's Tip

One way to distinguish between the three types of stationary point is to look at the sign of the derivative on either side of the point.

Mathematics Revision Notes

Differentiation

- At a local maximum the sign of the derivative changes from positive to negative.
- At a local minimum the sign of the derivative changes from negative to positive.
- At a stationary point of inflexion there is no change of sign.

The second derivative

Starting from $\frac{dy}{dx} = f'(x)$ and differentiating again gives the **second derivative**. This is written as $\frac{d^2y}{dx^2} = f''(x)$. The second derivative can give information about the nature of any stationary points.

At a stationary point:

- $f''(x) > 0 \Rightarrow$ the point is a local minimum
- $f''(x) < 0 \Rightarrow$ the point is a local maximum.

But, if $f''(x) = 0$ then this gives no further information.

Example Find the stationary points of the curve $y = \frac{x^3}{3} - \frac{3}{2}x^2 + 2x - 1$ and use the second derivative to distinguish between them.

First differentiate the equation, then solve $\frac{dy}{dx} = 0$ to locate the stationary points.

$\frac{dy}{dx} = x^2 - 3x + 2$. At a stationary point $\frac{dy}{dx} = 0 \Rightarrow x^2 - 3x + 2 = 0$
$\Rightarrow (x-1)(x-2) = 0$
$\Rightarrow x = 1$ or $x = 2$.

When $x = 1$, $y = \frac{1}{3} - \frac{3}{2} + 2 - 1 = -\frac{1}{6}$.

When $x = 2$, $y = \frac{8}{3} - 6 + 4 - 1 = -\frac{1}{3}$.

The stationary points are $(1, -\frac{1}{6})$ and $(2, -\frac{1}{3})$.

Using the second derivative: $f''(x) = 2x - 3$

so $f''(1) = -1 < 0$ giving a local maximum,

and $f''(2) = 1 > 0$ giving a local minimum.

Examiner's Tip

In this case, the function is a cubic and so the shape of the curve is known. We should expect the first stationary point to be a local maximum and the second to be a local minimum.

Integration

The idea of a **reverse process** is an important one in many areas of mathematics. In **calculus**, the reverse process of differentiation is **integration** and this turns out to be an extremely important process, in its own right, with many powerful applications.

When you differentiate x^n with respect to x you may think of the process involving two stages:

$$x^n \longrightarrow \boxed{\text{multiply by the power}} \longrightarrow \boxed{\text{reduce the power by 1}} \longrightarrow nx^{n-1}$$

This *suggests* that the reverse process is given by:

$$\frac{x^{n+1}}{n+1} \longleftarrow \boxed{\text{divide by the power}} \longleftarrow \boxed{\text{increase the power by 1}} \longleftarrow x^n$$

However, this doesn't give the *complete* picture. The reason is that you can differentiate $x^n +$ (any constant) and still obtain nx^{n-1}. You need to take this into account when you reverse the process.

The result is written as $\int x^n \, dx = \dfrac{x^{n+1}}{n+1} + c$ where c is called the **constant of integration**.

Some examples are:

$$\int x^2 \, dx = \frac{x^3}{3} + c \qquad \int \sqrt{x} \, dx = \int x^{\frac{1}{2}} \, dx = \frac{2}{3} x^{\frac{3}{2}} + c$$

$$\int \frac{1}{x^2} \, dx = \int x^{-2} \, dx = -x^{-1} + c = -\frac{1}{x} + c.$$

Sums and differences of functions are treated in the same way as in differentiation, by dealing with each term separately. The general rules are given below.

$$\int f(x) \pm g(x) \, dx = \int f(x) \, dx \pm \int g(x) \, dx \qquad \int af(x) \, dx = a \int f(x) \, dx \quad \text{(where } a \text{ is a constant)}$$

For example,

$$\int (3x^2 - 5x + 2) \, dx = x^3 - 5\frac{x^2}{2} + 2x + c.$$

Examiner's Tip

In some situations you have enough information to find the value of the constant. For example, you might be told the co-ordinates of a point on the curve.

Integration

Example Find the equation of the curve with gradient $3x^2$ passing through $(2, 5)$.

$$\frac{dy}{dx} = 3x^2 \Rightarrow y = \int 3x^2 \, dx \Rightarrow y = x^3 + c. \quad \text{When } x = 2, y = 5 \Rightarrow c = -3.$$

The equation of the curve is $y = x^3 - 3$.

The area under a curve

An integral can take different forms to suit a particular purpose. An integral of the form $\int f(x) \, dx = g(x)$ is called an **indefinite integral** and is useful when the end result is required to be a function. All of the ones looked at so far have been indefinite integrals.

An integral of the form $\int_a^b f(x) \, dx = [g(x)]_a^b$ is called a **definite integral** and has a numerical value given by $g(b) - g(a)$. This represents the area under the curve $y = f(x)$ between the lines $x = a$ and $x = b$.

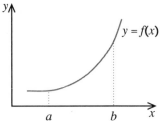

Example Find the area under the curve $y = x^2 + 1$ between $x = 1$ and $x = 3$.

$$\text{Area} = \int_1^3 (x^2 + 1) \, dx = \left[\frac{x^3}{3} + x\right]_1^3$$

$$= \left(\frac{3^3}{3} + 3\right) - \left(\frac{1^3}{3} + 1\right)$$

$$= 9 + 3 - \frac{1}{3} - 1$$

$$= 10\tfrac{2}{3}.$$

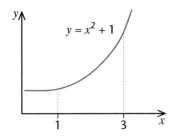

The area between a curve and the y-axis

The area bounded by the curve $y = f(x)$, the y-axis, and the lines $y = a$ and $y = b$ is given by $\int_a^b x \, dy$.

Example Find the area between the curve $y = x^3$ and the y-axis between $y = 1$ and $y = 8$.

Since $y = x^3$, it follows that $x = y^{\frac{1}{3}}$ and the required area is given by

$$\int_1^8 y^{\frac{1}{3}} \, dy = \left[\frac{3}{4} y^{\frac{4}{3}}\right]_1^8 = \frac{3}{4}(8^{\frac{4}{3}} - 1)$$

$$= \frac{3}{4} \times 15 = 11\tfrac{1}{4}.$$

Examiner's Tip

In this case, you have to write x in terms of y and integrate with respect to y.

Progress check

Algebra

1. A polynomial is given by $f(x) = 3x^3 + 11x^2 - 6x - 8$.
   - (a) Find the value of $f(-1)$ and $f(1)$.
   - (b) Factorise $f(x)$ completely.
   - (c) Solve $f(x) = 0$.
   - (d) Sketch the graph of $y = f(x)$.

2. (a) Solve $x^2 - 6x + 7 = 0$ by completing the square.
      Give the roots in surd form.
   (b) Solve $3x^2 + 11x - 2 = 0$ using the quadratic formula.
      Give the roots to 2 decimal places.

3. Find the values of A, B and C.
   $3x^3 - 16x - 6 \equiv Ax(x-1)(x+2) + B(x-1)^2 + C(4x^2 + 3)$.

Sequences and series

1. Write down the first five terms of the sequence given by:
   (a) $u_n = 2n^2$ (b) $u_1 = 10$, $u_{n+1} = 3u_n + 2$.

2. Find the sum of the first 1000 natural numbers.

3. Find the sum of the first 20 terms of $12 + 15 + 18.75 + \ldots$ to 4 s.f.

Trigonometry

1. Describe the transformations needed to turn the graph of $y = \cos x$ into the graph of: (a) $\cos 2x$ (b) $3 \cos x$ (c) $3 \cos(2x) - 1$.

2. Show that $(\sin x + \cos x)^2 + (\sin x - \cos x)^2 \equiv 2$.

3. Solve $8 \sin x = 3 \cos x$ for $-360° \leq x \leq 360°$.

Co-ordinate geometry

The equation of a straight line is $3x + 2y = 11$.
- (a) Find the equation of a parallel line crossing the point $(5,4)$.
- (b) Find the equation of a perpendicular line passing through the point $(-2,6)$.

Differentiation

1. Find the equation of: (a) the tangent (b) the normal
   to the curve $y = 12\sqrt{x}$ at the point $(9,36)$.

2. Locate the stationary points on the curve $y = 2x^3 - 6x^2 - 18x + 5$ and use the second derivative to distinguish between them.

Integration

1. Find the value of each of these definite integrals.
   (a) $\int_0^5 (x-2)\,dx$ (b) $\int_4^9 \sqrt{x}\,dx$ (c) $\int_0^6 y^2\,dy$

2. A curve has gradient function $4x^3 + 1$ and passes through the point $(1,9)$. Find the equation of the curve.

Answers on page 94

Pure 2

Algebra and functions

Composition of functions

The diagram shows how two functions f and g may be combined.

$$x \xrightarrow{f} f(x) \xrightarrow{g} g(f(x))$$

In this case, f is applied first to some value x giving $f(x)$. Then g is applied to the value $f(x)$ to give $g(f(x))$. This is usually written as $gf(x)$ and gf can be thought of as a new **composite function** defined from the functions f and g.

For example, if $f(x) = 3x$, $x \in \mathbb{R}$ and $g(x) = x + 2$, $x \in \mathbb{R}$
then $gf(x) = g(3x) = 3x + 2$.

The order in which the functions are applied is important. The composite function fg is found by applying g first and then f.

In this case, $fg(x) = f(x + 2) = 3x + 6$.

Generally speaking, when two functions f and g are defined, the composite functions fg and gf will not be the same.

The inverse of a function

The **inverse of a function** f is a function, usually written as f^{-1}, that *undoes* the effect of f. So the inverse of a function which adds 2 to every value, for example, will be a function that subtracts 2 from every value.

This can be written as $f(x) = x + 2$, $x \in \mathbb{R}$ and $f^{-1}(x) = x - 2$, $x \in \mathbb{R}$.
The domain of f^{-1} is given by the range of f.
Notice that $f^{-1}f(x) = f^{-1}(x + 2) = x$ and that $ff^{-1}(x) = f(x - 2) = x$.

A function can be either one–one or many–one, but only functions that are one–one can have an inverse. The reason is, that reversing a many–one function would give a mapping that is one–many, and this cannot be a function.

Examiner's Tip

Every value in the domain of a function can have only one image. This is an important property of functions.

The diagram shows a many–one function. It does not have an inverse.

If you reverse this diagram then p would have more than one image.

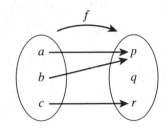

You can turn a many-one function into a one-one function by restricting its domain. For example, the function $f(x) = x^2$, $x \in \mathbb{R}$ is many-one and so it doesn't have an inverse. However, the function $f(x) = x^2$, $x > 0$ is one-one and $f^{-1}(x) = \sqrt{x}$, $x > 0$.

Finding the inverse of a one–one function

One way to find the inverse of a one-one function f is to write $y = f(x)$ and then rearrange this to make x the subject so that $x = f^{-1}(y)$.

The inverse function is then usually defined in terms of x to give $f^{-1}(x)$. The domain of f^{-1} is the range of f.

Example (a) Find the inverse of the function $f(x) = \dfrac{x+2}{x-3}$, $x \neq 3$.

(b) Sketch the graphs of $y = f(x)$ and $y = f^{-1}(x)$.

Start by writing $y = f(x)$

(a) Define $y = \dfrac{x+2}{x-3}$

then $y(x-3) = x + 2$

Collect the x terms on one side of the equation.

$xy - 3y = x + 2$

$xy - x = 3y + 2$

$x(y - 1) = 3y + 2$

Rearrange to find x

$x = \dfrac{3y+2}{y-1}$.

So $f^{-1}(x) = \dfrac{3x+2}{x-1}$, $x \neq 1$.

Examiner's Tip

It is usual to express the inverse function in terms of x.

Algebra and functions

The graph of the inverse function is the reflection of the original graph in the line $y = x$.

This is true for any inverse function but the same scale must be used on both axes or the effect is distorted.

Another point to notice is that the original graph approaches 1 but doesn't reach it so that 1 does not belong to the range of the function.

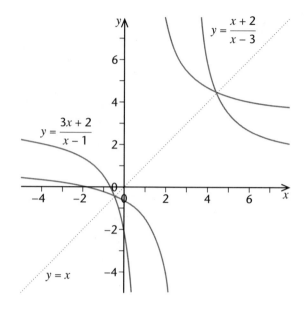

The modulus function

The notation $|x|$ is used to stand for the modulus of x. This is defined as

$$|x| = \begin{cases} x & \text{when } x \geq 0 \quad \text{(when } x \text{ is positive } |x| \text{ is just the same as } x\text{).} \\ -x & \text{when } x < 0 \quad \text{(when } x \text{ is negative } |x| \text{ is the same as } -x\text{).} \end{cases}$$

The graph of $y = |x|$

It follows that the graph of $y = |x|$ is the same as the graph of $y = x$ for positive values of x. But, when x is negative, the corresponding part of the graph of $y = x$ must be reflected in the x-axis to give the graph of $y = |x|$.

$|x|$ is never negative so the graph of $y = |x|$ doesn't go below the x-axis anywhere.

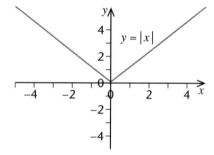

The graph of $y = |f(x)|$

The graph of $y = |f(x)|$ is the same as the graph of $y = f(x)$ for positive values of $f(x)$. But, when $f(x)$ is negative, the corresponding part of the graph of $y = f(x)$ must be reflected in the x-axis to give the graph of $y = |f(x)|$.

The diagram shows the graph of $y = |1/x|$.

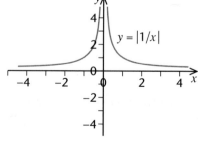

The graph of $y = f(|x|)$

The graph of $y = f(|x|)$ is the same as the graph of $y = f(x)$ when x is positive. This part of the graph is then reflected in the y-axis to give the graph of $y = f(|x|)$ for negative values of x.

The diagram shows the graphs of $y = (x-1)^2$ and $y = (|x|-1)^2$.

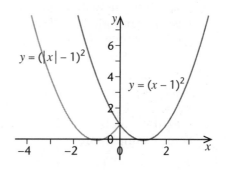

The expression $|x-a|$ can be interpreted as the distance between the numbers x and a on the number line. In this way, the statement $|x-a| < b$ means that the distance between x and a is less than b.

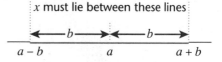

It follows that $a - b < x < a + b$.

Binomial expansion

An expression which has two terms, such as $a + b$ is called a **binomial**. The expansion of something of the form $(a+b)^n$ is called a **binomial expansion**. When n is a positive integer:

$$(a+b)^n = a^n + na^{n-1}b + \frac{n(n-1)}{2!}a^{n-2}b^2 + \frac{n(n-1)(n-2)}{3!}a^{n-3}b^3 + \ldots + b^n.$$

This expansion may appear complicated but it starts to look simpler if you follow the patterns from one term to the next:

- Starting with a^n, the power of a is reduced by 1 each time until the last term, b^n, which is the same as $a^0 b^n$.
- The power of b is increased by 1 each time. Notice that the first term, a^n, is the same as $a^n b^0$ and that the powers of a and b always add up to n in each term.

Examiner's Tip

It's worth remembering the first three coefficients: 1, n and $\frac{n(n-1)}{2!}$ then the rest follow the same pattern, giving $\frac{n(n-1)(n-2)}{3!}$, $\frac{n(n-1)(n-2)(n-3)}{4!}$ and so on.

Algebra and functions

The coefficients in the expansion of $(a + b)^n$ also follow the pattern given by the row of **Pascal's triangle** that starts with 1 n ...
So the coefficients in the expansion of $(a + b)^4$ for example are 1, 4, 6, 4 and 1.
It follows that
$(a + b)^4 = a^4 + 4a^3b + 6a^2b^2 + 4ab^3 + b^4$.

```
              1
           1     1
         1    2    1
       1    3    3    1
     1    4    6    4    1
   1    5   10   10    5    1
```

Variations on this result can be obtained by substituting different values for a and b. Some examples are:

$$(1 + x)^4 = 1 + 4x + 6x^2 + 4x^3 + x^4.$$

and

$$(1 - 2x)^4 = 1 + 4(-2x) + 6(-2x)^2 + 4(-2x)^3 + (-2x)^4$$
$$= 1 - 8x + 24x^2 - 32x^3 + 16x^4.$$

The notation $\binom{n}{r}$ is often used to stand for the expression $\dfrac{n(n-1)\ldots(n-r+1)}{r!}$.

So, for example $\binom{n}{3} = \dfrac{n(n-1)(n-2)}{3!}$.

Using this notation, the binomial expansion may be written as:

$$(a + b)^n = a^n + \binom{n}{1}a^{n-1}b + \binom{n}{2}a^{n-2}b^2 + \binom{n}{3}a^{n-3}b^3 + \ldots + b^n.$$

- It is useful to recognise that the term involving b^r takes the form $\binom{n}{r}a^{n-r}b^r$.

- For positive integer values $\binom{n}{r}$ has the same value as ^nC_r and you may find that your calculator will work this out for you.

- When n is small and the full binomial expansion is required, the simplest way to find the coefficients is to use Pascal's triangle. For larger values of n it may be simpler to use the formula, particularly if only some of the terms are required.

Example Expand $(1 + 3x)^{10}$ in ascending powers of x up to and including the fourth term.

$$(1 + 3x)^{10} = 1 + \binom{10}{1}(3x) + \binom{10}{2}(3x)^2 + \binom{10}{3}(3x)^3 + \ldots$$
$$= 1 + 10(3x) + 45(9x^2) + 120(27x^3) + \ldots$$
$$= 1 + 30x + 405x^2 + 3240x^3 + \ldots.$$

Example Find the coefficient of the x^7 term in the expansion of $(3 - 2x)^{15}$

The term involving x^7 is given by $\binom{15}{7}(3)^8(-2x)^7$

$= 6435 \times 6561 \times -128x^7$

and so the required coefficient is $-5\,404\,164\,480$.

Algebraic division

If you work out $27 \div 4$ for example, then you obtain 6 as the **quotient** and 3 as the **remainder**. You can write $27 = 4 \times 6 + 3$ to show the connection between these values.

The same thing applies in algebra when one polynomial is divided by another.

Example Find the quotient and remainder when $x^2 + 7x - 5$ is divided by $x - 4$.

Using $ax + b$ for the quotient and c for the remainder gives:

$x^2 + 7x - 5 \equiv (x - 4)(ax + b) + c.$ [1]
$\equiv ax^2 + (b - 4a)x + c - 4b.$

Equating coefficients of x^2 gives: $a = 1$.

Equating coefficients of x gives: $b - 4a = 7$
$\Rightarrow b - 4 = 7$
$\Rightarrow b = 11.$

Equating the constant terms gives: $c - 4b = -5$
$\Rightarrow c - 44 = -5$
$\Rightarrow c = 39.$

Substituting for a, b and c in [1] gives $x^2 + 7x - 5 \equiv (x - 4)(x + 11) + 39$.

So, the quotient is $(x + 11)$ and the remainder is 39.

The remainder theorem

In the previous example, the quotient and remainder of $(x^2 + 7x - 5) \div (x - 4)$ were found by using $x^2 + 7x - 5 \equiv (x - 4)(ax + b) + c$ and equating coefficients.

In its generalised form this result is known as the **remainder theorem** which states:

> When a polynomial $f(x)$ is divided by $(x - a)$ the remainder is $f(a)$.

Examiner's Tip

Another way to use this identity is to substitute particular values for x. Notice that when $x = 4$, the RHS simplifies to c and so the remainder is easily found by substituting $x = 4$ on the LHS. This gives $c = 4^2 + 7 \times 4 - 5 = 39$ as before.

Algebra and functions

Example Find the remainder when the polynomial $f(x) = x^3 + x - 5$ is divided by:
(a) $(x - 3)$ (b) $(x + 2)$ (c) $(2x - 1)$.

(a) The remainder is $f(3) = 3^3 + 3 - 5 = 25$.
(b) The remainder is $f(-2) = (-2)^3 - 2 - 5 = -15$.
(c) The remainder is $f(0.5) = 0.5^3 + 0.5 - 5 = -4.375$.

A special case of the remainder theorem occurs when the remainder is 0. This result is known as the **factor theorem** which states:

> If $f(x)$ is a polynomial and $f(a) = 0$ then $(x - a)$ is a factor of $f(x)$.

Example Show that $(x + 3)$ is a factor of $x^3 + 5x^2 + 5x - 3$.

Taking $f(x) = x^3 + 5x^2 + 5x - 3$

$f(-3) = (-3)^3 + 5(-3)^2 + 5(-3) - 3$

$= -27 + 45 - 15 - 3 = 0$.

By the factor theorem $(x + 3)$ is a factor of $x^3 + 5x^2 + 5x - 3$.

Exponential functions

An **exponential function** is one where the variable is a power or exponent. For example, any function of the form $f(x) = a^x$, where a is a constant, is an exponential function.

The diagram shows some graphs of exponential functions for different values of a.

All of the graphs pass through the point $(0,1)$ and each one has a different gradient at this point. The value of the gradient depends on a.

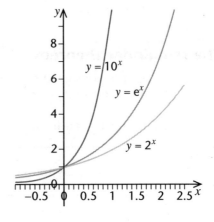

There is a particular value of a for which the curve has gradient 1 at $(0,1)$. The letter e is used to stand for this value and the function $f(x) = e^x$ is often called **the exponential function** because of its special importance to the subject.
$e = 2.718281828\ldots$ it is an irrational number and so it cannot be written exactly.

Logarithms

An exponential function such as $f(x) = 10^x$ is a one–one function and so it has an inverse. If $y = 10^x$ then the inverse function is called the **logarithm of y to base 10**. This function may be used to express x in terms of y. It is written as $x = \log_{10} y$. In the same way, if $y = e^x$ then $x = \log_e y$ but this is usually written as $x = \ln y$.

Logarithms to base e are called **natural logarithms** and are needed in integration.

Since the exponential and natural logarithm functions are inverses of each other, their graphs are symmetrical about the line $y = x$.

The diagram shows that the domain of the natural logarithm function is given by $x > 0$. This is true for logs to any base.

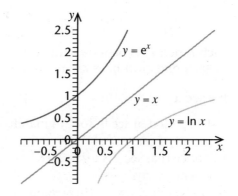

The log laws

There are three laws of logarithms that you need to know. The same laws apply in any base and so no particular base is stated:

- $\log a + \log b = \log ab$
- $\log a - \log b = \log\left(\dfrac{a}{b}\right)$
- $n \log a = \log a^n$.

You can use these laws to simplify expressions involving logs and to solve **exponential equations**, i.e. equations where the unknown value is a power.

Example 1   Express $\log x + 3 \log y$ as a single logarithm.

$$\log x + 3 \log y = \log x + \log y^3 \quad \text{(using the third law)}$$
$$= \log xy^3 \quad \text{(using the first law)}$$

Example 2   Solve the equation $5^x = 30$.

Taking logarithms of both sides gives:

$$\log 5^x = \log 30$$
$$\Rightarrow x \log 5 = \log 30$$

$$\Rightarrow x = \frac{\log 30}{\log 5} = 2.113 \text{ to 4 s.f.}$$

Examiner's Tip

Notice that $\log_a a = 1$ for any base a. In particular $\log_{10} 10 = 1$ and $\ln e = 1$.

Trigonometry

Secant, cosecant and cotangent

These trigonometric functions, commonly known as **sec**, **cosec** and **cot**, are defined from the more familiar sin, cos and tan functions as follows:

$$\sec x = \frac{1}{\cos x} \qquad \operatorname{cosec} x = \frac{1}{\sin x} \qquad \cot x = \frac{1}{\tan x} = \frac{\cos x}{\sin x}$$

Each function is periodic and has either line or rotational symmetry. The domain of each one must be restricted to avoid division by zero.

$y = \sec x$

- The period of sec is 360° to match the period of cos.
- Notice that sec x is undefined whenever cos $x = 0$.
- The graph is symmetrical about every vertical line passing through a vertex.
- It has rotational symmetry of order 2 about the points in the x-axis corresponding to $90° \pm 180°n$.

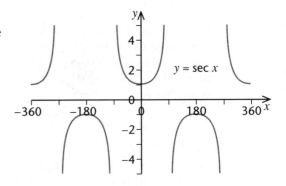

$y = \operatorname{cosec} x$

- The period of cosec is 360° to match the period of sin.
- Notice that cosec x is undefined whenever sin $x = 0$.
- The graph is symmetrical about every vertical line passing through a vertex.
- It has rotational symmetry of order 2 about the points on the x-axis corresponding to $0°$, $\pm 180°$, $\pm 360°$, ….

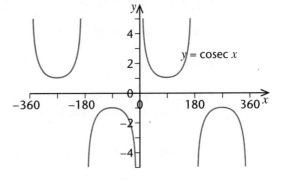

$y = \cot x$

- The period of cot is 180° to match the period of tan.
- Notice that cot x is undefined whenever sin $x = 0$.
- The graph has rotational symmetry of order 2 about the points on the x-axis corresponding to $0°$, $\pm 90°$, $\pm 180°$, ….

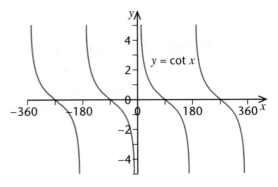

Inverse trigonometric functions

The sine, cosine and tangent functions are all many–one and so do not have inverses on their full domains. However, it is possible to restrict their domains so that each one has an inverse. The graphs of these inverse functions are given below.

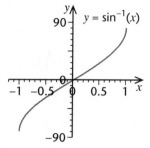
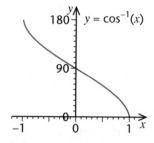
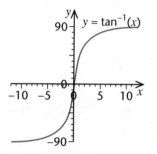

$f(x) = \sin x, -90° \leq x \leq 90°$
$\Rightarrow f^{-1}(x) = \sin^{-1} x, -1 \leq x \leq 1$.
$\sin^{-1} x$ means:
'the angle whose sine is x'
e.g. $\sin^{-1} 0.5 = 30°$.

$f(x) = \cos x, 0° \leq x \leq 180°$
$\Rightarrow f^{-1}(x) = \cos^{-1} x, -1 \leq x \leq 1$.
$\cos^{-1} x$ means:
'the angle whose cosine is x'
e.g. $\cos^{-1} 0.5 = 60°$.

$f(x) = \tan x, -90° < x < 90°$
$\Rightarrow f^{-1}(x) = \tan^{-1} x, x \in \mathbb{R}$.
$\tan^{-1} x$ means:
'the angle whose tangent is x'
e.g. $\tan^{-1} 1 = 45°$.

Trigonometric identities

If possible, you should *learn* the identities listed below. It is so much easier to simplify expressions, establish new identities and solve equations if you have a firm grasp of these basic results.

- The Pythagorean identities are:

  $\sin^2 \theta + \cos^2 \theta \equiv 1 \qquad 1 + \tan^2 \theta \equiv \sec^2 \theta \qquad \cot^2 \theta + 1 \equiv \text{cosec}^2 \theta$.

- The compound angle formulae for $(A + B)$ are:

  $$\sin(A + B) \equiv \sin A \cos B + \cos A \sin B$$
  $$\cos(A + B) \equiv \cos A \cos B - \sin A \sin B$$
  $$\tan(A + B) \equiv \frac{\tan A + \tan B}{1 - \tan A \tan B}.$$

The corresponding results for $(A - B)$ are:

$$\sin(A - B) \equiv \sin A \cos B - \cos A \sin B$$
$$\cos(A - B) \equiv \cos A \cos B + \sin A \sin B$$
$$\tan(A - B) \equiv \frac{\tan A - \tan B}{1 + \tan A \tan B}.$$

Examiner's Tip

These results can be worked out from the ones for $(A + B)$. Just replace every '+' with a '−' and vice versa. This makes it easier to learn the results.

Trigonometry

- The double angle formulae are:

$$\sin 2A \equiv 2 \sin A \cos A$$
$$\cos 2A \equiv \cos^2 A - \sin^2 A$$
$$\equiv 2\cos^2 A - 1$$
$$\equiv 1 - 2\sin^2 A$$
$$\tan 2A \equiv \frac{2 \tan A}{1 - 2\tan^2 A}.$$

- The half angle formulae are:

$$\sin^2 \tfrac{1}{2}A \equiv \tfrac{1}{2}(1 - \cos A)$$
$$\cos^2 \tfrac{1}{2}A \equiv \tfrac{1}{2}(1 + \cos A).$$

Proving identities

The basic technique for proving a given identity is:

- Start with the expression on one side of the identity.
- Use known identities to replace some part of it with an equivalent form.
- Simplify the result and compare with the expression on the other side.

With practice you get to know which parts to replace so that the required result is established.

Example

Prove that $\dfrac{\cos 2A}{\cos A - \sin A} \equiv \cos A + \sin A$.

Starting with the LHS:

$\dfrac{\cos 2A}{\cos A - \sin A} \equiv \dfrac{\cos^2 A - \sin^2 A}{\cos A - \sin A}$ using one of the double angle results.

$\equiv \dfrac{(\cos A + \sin A)(\cos A - \sin A)}{\cos A - \sin A}$ using the difference of two squares.

$\equiv \cos A + \sin A \equiv$ RHS.   cancelling the common factor.

Examiner's Tip

It is usually a good idea to replace sec, cosec and cot with $\dfrac{1}{\sin}$, $\dfrac{1}{\cos}$ and $\dfrac{1}{\tan}$ at the start.

Example

Prove that $\sec^2 A \equiv \dfrac{\operatorname{cosec} A}{\operatorname{cosec} A - \sin A}$.

$$\text{RHS} \equiv \dfrac{\dfrac{1}{\sin A}}{\dfrac{1}{\sin A} - \sin A} \equiv \dfrac{1}{1 - \sin^2 A}$$

Replacing $\operatorname{cosec} A$ with $\dfrac{1}{\sin A}$ and multiplying the numerator and denominator by $\sin A$.

$$\equiv \dfrac{1}{\cos^2 A} \equiv \sec^2 A \equiv \text{LHS}.$$

Using one of the Pythagorean identities and the definition of $\sec A$.

Solving equations

Identities may be used to re-write an equation in a form that is easier to solve.

Example Solve the equation $\cos 2x + \sin x + 1 = 0$ for $-\pi \leqslant x \leqslant \pi$.

Replacing $\cos 2x$ with $1 - 2\sin^2 x$ gives $1 - 2\sin^2 x + \sin x + 1 = 0$
$$\Rightarrow 2\sin^2 x - \sin x - 2 = 0.$$

Using the formula gives $\sin x = \dfrac{1 \pm \sqrt{1 + 16}}{4} = \dfrac{1 \pm \sqrt{17}}{4}$

$\Rightarrow \sin x = 1.28 \ldots$ (no solutions) or $\sin x = -0.78077 \ldots$.

This gives $\quad x = -0.8959$ or $-\pi + 0.8959 = -2.2457$ to 4 d.p.

An important result to be aware of is that the expression $a \cos \theta + b \sin \theta$ can be written in any of the equivalent forms $r \cos(\theta \pm \alpha)$ or $r \sin(\theta \pm \alpha)$ for a suitable choice of r and α. The convention is to take $r > 0$.

Example Express $3 \cos \theta - 4 \sin \theta$ in the form $r \cos(\theta + \alpha)$, $r > 0$.

$r \cos(\theta + \alpha) \equiv 3 \cos \theta - 4 \sin \theta$
$\Rightarrow r \cos \theta \cos \alpha - r \sin \theta \sin \alpha \equiv 3 \cos \theta - 4 \sin \theta$
$\Rightarrow r \cos \alpha = 3$ and $r \sin \alpha = 4$
$\Rightarrow r = \sqrt{3^2 + 4^2} = 5$ and $\alpha = \cos^{-1}(\tfrac{3}{5}) = 0.9273$ to 4 d.p.
$\Rightarrow 3 \cos \theta - 4 \sin \theta \equiv 5 \cos(\theta + 0.9273)$.

Examiner's Tip

One application of the result is to solve equations of the form $a \cos \theta + b \sin \theta = c$.

Differentiation

The exponential function e^x

The exponential function was defined on page 36. It has the special property that it is unchanged by differentiation. So, if $y = e^x$ then $\frac{dy}{dx} = e^x$

Constant multiples of the exponential function behave in the same way, so if $y = ae^x$ then $\frac{dy}{dx} = ae^x$ where a is any constant.

Examples

Differentiate with respect to x:

(a) $y = 2e^x$ (b) $y = x^3 + e^x$ (c) $f(x) = \frac{1}{x} - 3e^x$

(a) $\frac{dy}{dx} = 2e^x$ (b) $\frac{dy}{dx} = 3x^2 + e^x$ (c) $f'(x) = -\frac{1}{x^2} - 3e^x$

In general, if $y = ae^{kx}$ then $\frac{dy}{dx} = ake^{kx}$

For example, if $y = 3e^{5x}$ then $\frac{dy}{dx} = 3 \times 5e^{5x} = 15e^{5x}$.

The natural logarithm function $\ln x$

If $y = \ln x$ then $\frac{dy}{dx} = \frac{1}{x}$.

This result is easily extended to functions of the form $y = \ln kx$ using the first law of logarithms.

If $y = \ln kx$ (k is a constant)

then $y = \ln x + \ln k$

giving, $\frac{dy}{dx} = \frac{1}{x} + 0$ (since $\ln k$ is a constant, its derivative is zero)

so, $\frac{dy}{dx} = \frac{1}{x}$.

For example, if $y = \ln 5x + 3e^x$ then $\frac{dy}{dx} = \frac{1}{x} + 3e^x$

The chain rule

The **chain rule** can be written as $\dfrac{dy}{dx} = \dfrac{dy}{du} \times \dfrac{du}{dx}$.

It shows how to differentiate a composite function by separating the process into simpler stages and combining the results.

Example Differentiate (a) $y = (3x + 2)^{10}$
(b) $y = \sqrt{x^2 - 9}$.

(a) Taking $u = 3x + 2$ gives $y = u^{10}$ and $\dfrac{dy}{du} = 10u^9$, so $\dfrac{dy}{du} = 10(3x + 2)^9$.

It also gives $\dfrac{du}{dx} = 3$.

So, from the chain rule: $\dfrac{dy}{dx} = 10(3x + 2)^9 \times 3 = 30(3x + 2)^9$.

(b) Taking $u = x^2 - 9$ gives $y = \sqrt{u} = u^{\frac{1}{2}}$, and $\dfrac{dy}{du} = \tfrac{1}{2} u^{-\frac{1}{2}}$, so $\dfrac{dy}{du} = \tfrac{1}{2}(x^2 - 9)^{-\frac{1}{2}}$.

It also gives $\dfrac{du}{dx} = 2x$.

So, from the chain rule: $\dfrac{dy}{dx} = \tfrac{1}{2}(x^2 - 9)^{-\frac{1}{2}} \times 2x = \dfrac{x}{\sqrt{x^2 - 9}}$.

The chain rule can also be used to establish results for **connected rates of change**.

For example, the rate of change of the volume of a sphere can be written as $\dfrac{dv}{dt}$

The corresponding rate of change of the radius of the sphere can be written as $\dfrac{dr}{dt}$.

Using the chain rule, the connection between these rates of change is given by:

$$\dfrac{dv}{dt} = \dfrac{dv}{dr} \times \dfrac{dr}{dt}.$$

Also, since $v = \tfrac{4}{3}\pi r^3$, it follows that $\dfrac{dv}{dr} = 4\pi r^2$,

so, the connection can now be written as $\dfrac{dv}{dt} = 4\pi r^2 \dfrac{dr}{dt}$.

Examiner's Tip

Practise using the chain rule until you are really comfortable with it. It is one of the most important results in calculus.

Integration

The reciprocal function $\dfrac{1}{x}$

The result $\displaystyle\int x^n\, dx = \dfrac{x^{n+1}}{n+1} + c$ holds for any value of n other than $n = -1$.

In this special case, the result takes a different form:

$$\int \dfrac{1}{x}\, dx = \ln|x| + c.$$

> The result breaks down with $n = -1$ because division by zero is undefined.

This extends to $\displaystyle\int \dfrac{k}{x}\, dx = k\int \dfrac{1}{x}\, dx = k\ln|x| + c,$

and to $\displaystyle\int \dfrac{1}{kx}\, dx = \dfrac{1}{k}\int \dfrac{1}{x}\, dx = \dfrac{1}{k}\ln|x| + c.$ (where k is a constant)

For example, $\displaystyle\int \dfrac{2}{x}\, dx = 2\ln|x| + c$ and $\displaystyle\int \dfrac{1}{3x}\, dx = \dfrac{1}{3}\ln|x| + c.$

You might need to do some algebraic manipulation first, to put a function in the right form to use these results.

Example Integrate $\dfrac{x+3}{x}$ with respect to x.

$$\int \dfrac{x+3}{x}\, dx = \int 1 + \dfrac{3}{x}\, dx$$
$$= x + 3\ln|x| + c.$$

The exponential function e^x

The fact that differentiation leaves the exponential function unchanged was stated on page 42. Since integration is the reverse process of differentiation it follows that

$$\int e^x\, dx = e^x + c \quad \text{and} \quad \int ae^x\, dx = a\int e^x\, dx = ae^x + c.$$

An extension of this result is $\displaystyle\int e^{kx}\, dx = \dfrac{1}{k}e^{kx} + c$, where k is a constant.

Examples

1. $\displaystyle\int 4e^x\, dx = 4e^x + c$
2. $\displaystyle\int \dfrac{1}{\sqrt{3}} e^x\, dx = \dfrac{1}{\sqrt{3}} e^x + c$

3. $\displaystyle\int e^{\frac{x}{2}}\, dx = \dfrac{1}{\left(\frac{1}{2}\right)} e^{\frac{x}{2}} + c = 2e^{\frac{x}{2}} + c.$

Volume of revolution

The shaded region R is bounded by the curve $y = f(x)$, the x-axis and the lines $x = a$ and $x = b$.

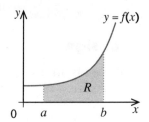

Rotating R completely about the x-axis forms a solid figure. The volume of this figure is called the volume of revolution and is given by

$$V = \int_a^b \pi y^2 \, dx$$

The corresponding result for rotation about the y-axis is

$$\int_a^b \pi x^2 \, dy$$

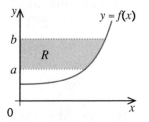

where the region is bounded by the curve $y = f(x)$, the y-axis and the lines $y = a$ and $y = b$.

Example Find a formula for the volume of a cone of height h and base radius r.

The straight line has gradient $\dfrac{r}{h}$ and passes through the origin, so its equation is $y = \dfrac{r}{h} x$.

The volume of the cone is given by the volume of revolution of the shaded region shown.

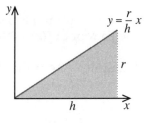

Using the formula for volume of revolution gives:

$$V = \int_0^h \pi \left(\dfrac{r}{h} x\right)^2 dx = \dfrac{\pi r^2}{h^2} \int_0^h x^2 \, dx$$

$$= \dfrac{\pi r^2}{h^2} \left[\dfrac{x^3}{3}\right]_0^h = \dfrac{\pi r^2}{h^2} \times \dfrac{h^3}{3}$$

$$= \dfrac{\pi r^2 h}{3}.$$

Examiner's Tip

You may be asked to find a volume of revolution about either axis in the exam. Make sure that you know which formula to apply in each case.

Numerical methods

Change of sign

If there is an interval from $x = a$ to $x = b$ in which the graph of a function has no breaks then the function is said to be **continuous** on the interval.

For example, the graph of $y = x^2 - 3$ is continuous everywhere.

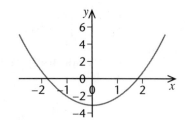

However, the graph of $y = \dfrac{1}{x - 3}$ is not continuous on any interval that contains the value $x = 3$.

All polynomial functions are continuous everywhere.

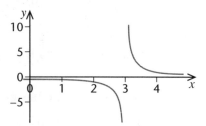

If $f(x)$ is continuous between $x = a$ and $x = b$ and if $f(a)$ and $f(b)$ have different signs, then a root of $f(x) = 0$ lies in the interval from a to b.

Example Given that $f(x) = e^x - 10x$, show that the equation $f(x) = 0$ has a root between 3 and 4.

$f(3) = -9.914\ldots < 0$

$f(4) = 14.598\ldots > 0$.

The change of sign shows that $f(x) = 0$ has a root between 3 and 4.

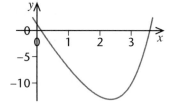

You can continue this process, called a **decimal search**, to trap the root in a smaller and smaller interval. The table gives the values of the function in steps of 0.1.

x	3.1	3.2	3.3	3.4	3.5	3.6	3.7	3.8	3.9
$f(x)$	−8.802	−7.467	−5.887	−4.035	−1.884	0.5982	3.4473	6.7011	10.402

This shows that the root lies between 3.5 and 3.6. At the mid-point $f(3.55) = -0.6866\ldots < 0$ and so the root lies between 3.55 and 3.6.

It follows that the root is nearer to 3.6 than 3.5 so the root is 3.6 to 1 d.p.

Iteration

To solve an equation by **iteration** you start with some approximation to a root and improve its accuracy by substituting it into a formula. You can then repeat the process until you have the desired level of accuracy.

Using **simple iteration**, the first step is to rearrange the equation to express x as a function of itself. This function defines the iterative formula that you need.

For example, the equation $e^x - 10x = 0$ can be rearranged as
$$e^x = 10x$$
$$\Rightarrow x = \ln(10x).$$

This may now be turned into an iterative formula for finding x.
$$x_{n+1} = \ln(10x_n).$$

Starting with $x_1 = 3$,
$x_2 = \ln 30 = 3.401\ldots$ and so on. This produces a sequence of values:

$x_2 = 3.401\ldots$ $x_6 = 3.5760\ldots$
$x_3 = 3.526\ldots$ $x_7 = 3.5768\ldots$
$x_4 = 3.563\ldots$ $x_8 = 3.5770\ldots$
$x_5 = 3.573\ldots$ $x_9 = 3.5771\ldots$

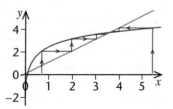

The calculator display soon settles on 3.577152064.
The value of the root is $x = 3.577$ to 4 s.f.

Start from a point on the x-axis, move up to the curve and across to $y = x$. Then move up to the curve and across to $y = x$ again. Continue in the same way. Each of these stages corresponds to one iteration of the formula.

The diagram shows how the process converges from a starting point on either side of the root.

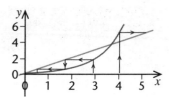

A different rearrangement of the original equation gives $x = \dfrac{e^x}{10}$ and so the corresponding iterative formula is $x_{n+1} = \dfrac{e^{x_n}}{10}$.

Notice how the movement is away from the upper root this time. If the starting value is smaller than the upper root then the process converges to the lower root.
A starting value above the upper root fails to converge to either root.

Starting with $x = 3$, this rearrangement fails to converge to the root between 3 and 4 but it does converge to the other root of the equation.

The value of this root is $x = 0.1118$ to 4 s.f.

Examiner's Tip

In some exam questions you may be given the iteration formula to start with.

Numerical methods

Numerical integration

The value of $A = \int_a^b f(x)\,dx$ represents the area under the graph of $y = f(x)$ between $x = a$ and $x = b$. You can find an approximation to this value using the **trapezium rule**. This is especially useful if the function is difficult to integrate.

The area under the curve between a and b may be divided into n strips of equal width d.

It follows that $d = \dfrac{b-a}{n}$.

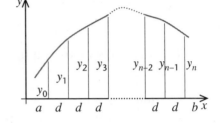

Each strip is approximately a trapezium and so the total area is approximately

$$\tfrac{1}{2}d(y_0+y_1) + \tfrac{1}{2}d(y_1+y_2) + \tfrac{1}{2}d(y_2+y_3) + \ldots + \tfrac{1}{2}d(y_{n-2}+y_{n-1}) + \tfrac{1}{2}d(y_{n-1}+y_n).$$

This simplifies to give the formula known as the trapezium rule:

$$A \approx \tfrac{1}{2}d(y_0 + 2(y_1 + y_2 + y_3 + \ldots y_{n-1}) + y_n).$$

Example

Use the trapezium rule with five strips to estimate the value of $\int_1^2 2^x\,dx$.

It is useful to tabulate the information:

x	1	1.2	1.4	1.6	1.8	2
2^x	2	2.2974	2.6390	3.0314	3.4822	4
	y_0	y_1	y_2	y_3	y_4	y_5

$$d = \dfrac{2-1}{5} = 0.2$$

So, $\int_1^2 2^x\,dx \approx 0.1\,(2 + 2(2.2974 + 2.6390 + 3.0314 + 3.4822) + 4)$

$\phantom{\text{So, }\int_1^2 2^x\,dx } = 2.89$ to 3 s.f.

Examiner's Tip

In this case, the working was done to 5 s.f. and the final answer is given to 3 s.f. It is a good idea to keep two extra figures in your working if possible.

Progress check

Algebra and functions

1. $f(x) = x^2 + 5$ and $g(x) = 2x + 3$.
   (a) Write $fg(x)$ in terms of x.
   (b) Find $fg(10)$.
   (c) Find the values of x for which $fg(x) = gf(x)$.

2. Expand $(1 + 3x)^{12}$ in ascending powers of x up to and including the third term.

3. Solve the equation $4^x = 100$ and give your answer to 4 s.f.

Pure 2 Trig

1. Prove the identity $\sin 2\theta \equiv \dfrac{2\tan\theta}{1+\tan^2\theta}$.

2. Solve the equation $\cos 2x - \sin x = 0$ for $0° \leqslant x \leqslant 360°$.

3. (a) Write $5\sin x + 12\cos x$ in the form $r\sin(x + \alpha)$ where $0 < \alpha < \dfrac{\pi}{2}$.
   (b) Solve the equation $5\sin x + 12\cos x = 7$ for $-\pi \leqslant x \leqslant \pi$.

Differentiation

Find $\dfrac{dy}{dx}$ in each of the following questions.

1. (a) $y = 10e^x$ (b) $y = 4e^x - x^3$ (c) $y = 4e^{2x}$

2. (a) $y = \ln 2x$ (b) $y = \ln 7x - 3e^x$ (c) $y = (1 - \sqrt{2x})^{11}$

Integration

1. Integrate with respect to x.
   (a) $\dfrac{3}{x}$ (b) $\dfrac{x+3}{x^2}$ c) $3e^x - \dfrac{1}{x}$.

2. The region bounded by the curve $y = e^{x/2}$, the lines $x = 1$ and $x = 3$ and the x-axis is rotated completely about the x-axis. Find the volume of revolution.

Numerical methods

1. (a) Show that the equation $x^3 - x - 7 = 0$ has a root between 2 and 3.
   (b) Use a decimal search to find the value of the root to 2 d.p.
   (c) Use simple iteration to find the value of the root to 4 d.p.

2. Use the trapezium rule with five strips to find the value of $\displaystyle\int_2^3 x\ln x\, dx$ to 2 d.p.

Answers on page 94

Mechanics 1

Vectors

A **scalar** quantity has size (or **magnitude**) but not direction. **Numbers** are scalars and some other important examples are **distance**, **speed**, **mass** and **time**.

A **vector** quantity has both size and **direction**. For example, distance in a specified direction is called **displacement**. Some other important examples are **velocity**, **acceleration**, **force** and **momentum**.

The diagram shows a **directed line segment**. It has size (in this case, length) and direction so it is a vector.

The diagram gives a useful way to represent *any* vector quantity and may also be used to represent addition and subtraction of vectors and multiplication of a vector by a scalar.

In a textbook, vectors are usually labelled with lower case letters in **bold** print.
When hand-written, these letter should be underlined e.g. $\underline{a}$

Addition and subtraction of vectors

This diagram shows three vectors **a**, **b** and **c** such that **c** = **a** + **b**.

The vectors **a** and **b** follow on from each other and then **c** joins the start of vector **a** to the end of vector **b**.

The vector **c** is called the **resultant** of **a** and **b**.

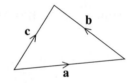

On a vector diagram, −**q** has the opposite sense of direction to **q**.
Notice that **p** − **q** is represented as **p** + (−**q**).

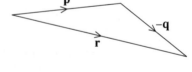

This diagram shows the same information in a different way.

Following the route in the opposite direction to **q** is the same as adding −**q** or subtracting **q**.

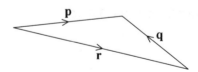

Scalar multiplication

2**p** is parallel to **p** and has the same sense of direction but is twice as long.

−3**p** is parallel to **p** but has the opposite sense of direction and is three times as long.

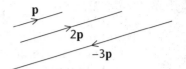

Component form

When working in two dimensions, it is often very useful to express a vector in terms of two special vectors **i** and **j**. These are **unit vectors** at right-angles to each other.
A vector **r** written as **r** = a**i** + b**j** is said to have **components** a**i** and b**j**.

For work involving the Cartesian co-ordinate system, **i** and **j** are taken to be in the positive directions of the x- and y-axes respectively.

In three dimensions, a third vector **k** is used to represent a unit vector in the positive direction of the z-axis.

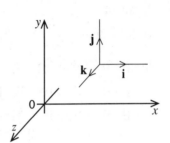

The **position vector** of a point P is the vector $\overrightarrow{OP}$ where O is the origin.
If P has co-ordinates (a, b) then its position vector is given by **r** = a**i** + b**j**.

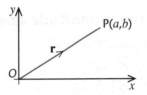

In three dimensions, a point with co-ordinates (a, b, c) would have position vector
r = a**i** + b**j** + c**k**.

Resolving a vector

The magnitude of **r** may be written as $|\mathbf{r}|$ or simply as r.

From the diagram, **r** = $r \cos \theta$**i** + $r \sin \theta$**j**.

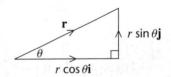

Examiner's Tip

If you know the magnitude and direction of a vector then you can use trigonometry to **resolve** it in terms of **i** and **j** components.

Vectors

Example
A force **f** has magnitude 10 N at an angle of 60° above the horizontal. Express **f** in the form $a\mathbf{i} + b\mathbf{j}$ where **i** and **j** are unit vectors in the horizontal and vertical directions respectively.

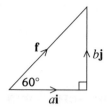

In this case, $a = 10 \cos 60° = 5$ and $b = 10 \sin 60° = 5\sqrt{3}$ giving $\mathbf{f} = 5\mathbf{i} + 5\sqrt{3}\mathbf{j}$.

Adding and subtracting vectors in component form

One advantage of expressing vectors in terms of **i** and **j** is that addition and subtraction may be done algebraically.

Example

$\mathbf{p} = 3\mathbf{i} + 2\mathbf{j}$ and $\mathbf{q} = 5\mathbf{i} - \mathbf{j}$. Express the following vectors in terms of **i** and **j**:

(a) $\mathbf{p} + \mathbf{q}$ (b) $\mathbf{p} - \mathbf{q}$ (c) $2\mathbf{p} - 3\mathbf{q}$.

(a) $\mathbf{p} + \mathbf{q} = (3\mathbf{i} + 2\mathbf{j}) + (5\mathbf{i} - \mathbf{j}) = 8\mathbf{i} + \mathbf{j}$
(b) $\mathbf{p} - \mathbf{q} = (3\mathbf{i} + 2\mathbf{j}) - (5\mathbf{i} - \mathbf{j}) = -2\mathbf{i} + 3\mathbf{j}$
(c) $2\mathbf{p} - 3\mathbf{q} = 2(3\mathbf{i} + 2\mathbf{j}) - 3(5\mathbf{i} - \mathbf{j}) = 6\mathbf{i} + 4\mathbf{j} - 15\mathbf{i} + 3\mathbf{j} = -9\mathbf{i} + 7\mathbf{j}$.

Finding the magnitude and direction of a vector

Another advantage of expressing a vector in terms of **i** and **j** components is that it is easy to find its magnitude and direction.

If $\mathbf{r} = a\mathbf{i} + b\mathbf{j}$ then the magnitude of **r** is given by

$$|\mathbf{r}| = \sqrt{a^2 + b^2}.$$

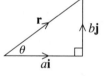

And the direction of **r** relative to **i** is given by

$$\theta = \tan^{-1}\left(\frac{b}{a}\right).$$

For example, if $\mathbf{r} = 3\mathbf{i} + 4\mathbf{j}$ then

$$|\mathbf{r}| = \sqrt{3^2 + 4^2} = 5 \text{ and } \theta = \tan^{-1}(\tfrac{4}{3}) = 53.1°.$$

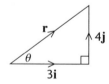

Examiner's Tip

It is always useful to draw a diagram so that you have a visual check for your answers.

Kinematics

Motion in a straight line

You can use these formulae whenever the acceleration of an object is constant.

$v = u + at$ s is the displacement from a fixed position

$s = ut + \frac{1}{2}at^2$ u is the initial velocity

$s = \left(\dfrac{u+v}{2}\right)t$ a is the acceleration

t is the time that the object has been in motion

$v^2 = u^2 + 2as$ v is the velocity at time t.

Example

An object starts from rest and moves in a straight line with constant acceleration 3 m s^{-2}. Find its velocity after 5 s.

From the given information: $u = 0$
$a = 3$
$t = 5$.

Don't include units in your *working*.

Using $v = u + at$
gives $v = 0 + 3 \times 5 = 15$.

Make a clear statement and include the appropriate units.

The velocity after 5 s is 15 m s^{-1}.

Example

A stone is thrown vertically upwards with a velocity of 20 m s^{-1}. It has a downward acceleration due to gravity of 10 m s^{-2}. Find the greatest height reached by the stone.

From the given information: $u = 20$
$a = -10$.

At the greatest height $v = 0$.
Using $v = u + at$
              $0 = 20 - 10t \Rightarrow t = 2$.
Using $s = ut + \frac{1}{2}at^2$
gives $s = 20 \times 2 - 5 \times 4 = 20$

The greatest height reached is 20 m.

Examiner's Tip

Motion may take place in either direction along a straight line. One direction is taken to be positive for displacement, velocity and acceleration and the other direction is taken to be negative.

Kinematics

Graphical representation

You need to know the properties of the graphs of distance, displacement, speed, velocity and acceleration against time.

- The gradient of a distance/time graph represents speed.
- The gradient of a displacement/time graph represents velocity.
- The gradient of a velocity/time graph represents acceleration.

- The area under a speed/time graph represents the distance travelled.
- The area under a velocity/time graph represents the change of displacement.
- The area under an acceleration/time graph represents change in velocity.

Example

The diagram represents the progress of a car as it travels between two sets of traffic lights.

Find:

(a) The initial acceleration of the car.

(b) The deceleration of the car as it approaches the second set of lights.

(c) The distance between the traffic lights.

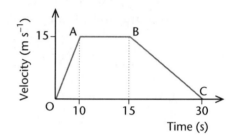

(a) Gradient of OA = $\frac{15}{10}$ = 1.5.

The initial *acceleration* of the car is 1.5 m s^{-2}.

(b) Gradient of BC = $-\frac{15}{15}$ = -1.

The *acceleration* of the car is -1 m s^{-2} so the *deceleration* 1 m s^{-2}

(c) The area of the trapezium OABC is given by $\frac{15}{2}(5 + 30) = 262.5$, so, the distance between the traffic lights is 262.5 m.

Variable acceleration

In this case, the constant acceleration formulae do not apply and must not be used. Using x for displacement, v for velocity and a for acceleration:

$$\begin{array}{c} \xrightarrow{\text{differentiate}} \\ x \quad v \quad a \\ \xleftarrow{\text{integrate}} \end{array} \qquad v = \frac{dx}{dt} \quad a = \frac{dv}{dt} = \frac{d^2x}{dt^2}$$

$$x = \int v\,dt \qquad v = \int a\,dt.$$

Example

A particle P moves along the x-axis such that its velocity at time t is given by $v = 5t - t^2$. When $t = 0$, $x = 15$. Find a formula for:

(a) the acceleration of the particle at time t
(b) the position of the particle at time t.

(a) $v = 5t - t^2 \Rightarrow \dfrac{dv}{dt} = 5 - 2t$.

The acceleration of the particle at time t is given by $a = 5 - 2t$.

(b) $x = \displaystyle\int v\,dt = \int 5t - t^2\,dt$

$\Rightarrow x = \dfrac{5t^2}{2} - \dfrac{t^3}{3} + c.$

When $t = 0$, $x = 15$

so $15 = 0 + c \Rightarrow c = 15$.

This gives the position of the particle at time t as $x = \dfrac{5t^2}{2} - \dfrac{t^3}{3} + 15$.

Using vectors

The work of this section may be extended to 2 and 3 dimensions using vectors.

Example

A particle has velocity $2\mathbf{i} - 3\mathbf{j}$ m s^{-1} when $t = 0$ and a constant acceleration of $\mathbf{i} + \mathbf{j}$ m s^{-2}. Find the speed of the particle when $t = 5$ and give its direction.

From the given information: $\mathbf{u} = 2\mathbf{i} - 3\mathbf{j}$
$\mathbf{a} = \mathbf{i} + \mathbf{j}$
$t = 5$.

Using $\mathbf{v} = \mathbf{u} + \mathbf{a}t$ gives $\mathbf{v} = 2\mathbf{i} - 3\mathbf{j} + 5(\mathbf{i} + \mathbf{j}) = 7\mathbf{i} + 2\mathbf{j}$.

Speed is the magnitude of the velocity,

giving $v = \sqrt{7^2 + 2^2} = 7.28\ldots$.

So the speed of the particle when $t = 5$ is 7.28 m s^{-1} to 2 d.p.

From the diagram, $\tan\theta = \tfrac{2}{7} \Rightarrow \theta = 15.9°$.

When $t = 5$ the particle is moving at 15.9° ↺ to the direction of $\mathbf{i}$.

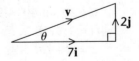

Examiner's Tip

The constant acceleration formulae work equally well in 2 or 3 dimensions.

AS Mathematics Revision Notes

Kinematics

Variable acceleration

Example

The position vector of a particle at time t is given by $\mathbf{r} = t^3 \mathbf{i} + 6t\mathbf{j}$. Find the velocity and acceleration of the particle at time t.

The velocity at time t is given by
$$\mathbf{v} = \frac{d\mathbf{r}}{dt} = \frac{d(t^3)}{dt}\mathbf{i} + \frac{d(6t)}{dt}\mathbf{j}$$
$$= 3t^2 \mathbf{i} + 6\mathbf{j}.$$

The acceleration at time t is given by
$$\mathbf{a} = \frac{d\mathbf{v}}{dt} = 6t\mathbf{i}.$$

Projectiles

The approach used for many projectile problems is to resolve the initial velocity into horizontal and vertical components and to treat these two parts separately.

Example

A stone is thrown horizontally with a speed of 10 m s^{-1} from the top of a cliff. The cliff is 98 m high and the stone lands in the sea d m from its base. Take $g = 9.8$ m s^{-2} and show that $d = 20\sqrt{5}$.

Vertically: $u = 0$
$a = 9.8$
$s = 98$.

Using $s = ut + \tfrac{1}{2}at^2$
gives $98 = 4.9t^2$
$\Rightarrow t = \sqrt{20} = 2\sqrt{5}$.

Horizontally: $d = 10 \times 2\sqrt{5}$
$\Rightarrow d = 20\sqrt{5}$.

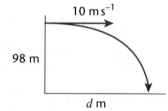

The initial velocity of the projectile may be at an angle θ to the horizontal.

Example

A ball is struck with velocity 20 m s^{-1} at 40° to the horizontal from a point 1 m above the ground. Find the maximum height reached by the ball. Take $g = 9.8$ m s^{-2}.

Vertically: $u = 20 \sin 40°$
$a = -9.8$.
At max ht: $v = 0$.
Using $v = u + at$
$0 = 20 \sin 40° - 9.8t$
$\Rightarrow t = 1.3118\ldots$.
Using $s = ut + \tfrac{1}{2}at^2$.
At max ht $s = 20 \sin 40° \times 1.3118 - 4.9 \times 1.3118^2 = 8.432\ldots$.

The maximum height is 9.43 m to 2 d.p.

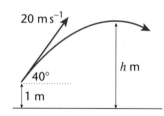

Examiner's Tip

Always draw a diagram to represent the information.

Statics

Force

Force is a vector quantity that influences the motion of an object. It is measured in newtons (N). For example, the **weight** of an object is the force exerted on it by gravity. An object with a mass of m kg has weight mg N. **Tension**, **reaction** and **friction** are other examples of force and will be met in this section.

Resolving forces

Here are some examples of resolving forces into two components at right-angles to each other.

In each case the forces are represented in magnitude and direction by the sides of the triangle. The force to be resolved is *always* shown on the hypotenuse.

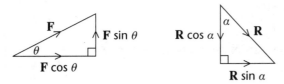

These are **vector diagrams** showing the relationship between a force and its components.

In a **force diagram**, you can *replace* a force with a pair of components that are equivalent to it. This will often make it easier to produce the equations necessary to solve a problem.

An important example is an object on an **inclined plane**.

The diagram shows an object of weight W on a smooth plane inclined at angle θ to the horizontal.

R is the force that the plane exerts on the object. It acts at right-angles to the plane and is called the **normal reaction**.

This vector diagram shows the component of weight acting down the plane is $W \sin \theta$ and the component perpendicular to the plane is $W \cos \theta$.

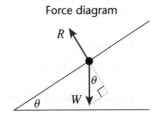

Force diagram

Vector diagram

Examiner's Tip

Take care not to show both the force *and* its components on a force **diagram** or you will represent the force TWICE.

AS Mathematics Revision Notes

Statics

Resultant force

The effect of several forces acting at a point is the same as the effect of a single force called the **resultant force**.

You can find the resultant of a set of forces by using vector addition.

Example

Find the magnitude and direction of the resultant $\mathbf{R}$ N of the forces $(3\mathbf{i} - 2\mathbf{j})$ N, $(\mathbf{i} + 3\mathbf{j})$ N and $(-2\mathbf{i} + 4\mathbf{j})$ N.

The resultant force is given by: $\mathbf{R} = (3\mathbf{i} - 2\mathbf{j}) + (\mathbf{i} + 3\mathbf{j}) + (-2\mathbf{i} + 4\mathbf{j}) = 2\mathbf{i} + 5\mathbf{j}$.

The magnitude of the resultant is $|\mathbf{R}| = \sqrt{2^2 + 5^2} = \sqrt{29}$.

From the diagram: $\theta = \tan^{-1}\left(\frac{5}{2}\right) = 68.2°$.

The resultant has magnitude $\sqrt{29}$ N and acts at 68.2° to the direction of $\mathbf{i}$.

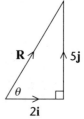

Equilibrium

A set of forces acting at a point is in **equilibrium** if the resultant force is zero.

It follows that the resolved parts of the forces must balance in any chosen direction.

Example

Find the force $\mathbf{F}$ such that the forces $(5\mathbf{i} - 3\mathbf{j})$, $(2\mathbf{i} + \mathbf{j})$ and $\mathbf{F}$ are in equilibrium.

For equilibrium $(5\mathbf{i} - 3\mathbf{j}) + (2\mathbf{i} + \mathbf{j}) + \mathbf{F} = 0$

$\Rightarrow 7\mathbf{i} - 2\mathbf{j} + \mathbf{F} = 0$

$\Rightarrow \mathbf{F} = -7\mathbf{i} + 2\mathbf{j}$.

Friction

Whenever two rough surfaces are in contact, the tendency of either surface to move relative to the other is opposed by the **force of friction**.

The diagram shows an object of weight W N resting on a rough horizontal surface.
The object is pushed from one side by a force P N and friction responds with force F N in the opposite direction.

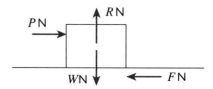

For small values of P, no movement takes place and $F = P$.

If P increases then F increases to maintain equilibrium until F reaches a maximum value. At this point the object is in **limiting equilibrium** and is on the point of slipping.

If P is now increased again then F will remain the same. The equilibrium will be broken and the object will move.

The maximum value of F depends on:

- the magnitude of the normal reaction R
- the roughness of the two surfaces measured by the value μ. This value is known as the **coefficient of friction**.

In general: $\quad\quad\quad\quad\quad F \leq \mu R$.
For limiting equilibrium: $\quad F = \mu R$.

Example

An object of mass 8 kg rests on a rough horizontal surface. The coefficient of friction is 0.3 and a horizontal force P N acts on the object which is about to slide. Take $g = 10 \text{ m s}^{-2}$ and find the value of P.

Resolving vertically gives $R = 80$.

Friction is limiting so $F = \mu R$
$= 0.3 \times 80 = 24$.

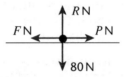

Resolving horizontally: $P = F \Rightarrow P = 24$.

Example

In the diagram, the object of mass m kg is on the point of slipping down the plane. The coefficient of friction between the object and the plane is μ. Show that $\mu = \tan \theta$.

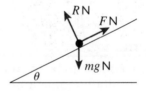

Resolving perpendicular to the plane $\quad R = mg \cos \theta$.
Resolving parallel to the plane $\quad\quad\quad F = mg \sin \theta$.
Friction is limiting so $\quad\quad\quad\quad\quad\quad F = \mu R$.
This gives $\quad\quad\quad\quad\quad\quad\quad\quad mg \sin \theta = \mu mg \cos \theta$

$$\Rightarrow \mu = \frac{\sin \theta}{\cos \theta} = \tan \theta.$$

Examiner's Tip

State which directions you choose to resolve in so that your method is clear.

Statics

Moments

The **moment of a force** about a point is a measure of the turning effect of the force about the point.
It is found by multiplying the force by its perpendicular distance from the point.
The moment of a force acts either clockwise ↻ or anticlockwise ↺ about the point.

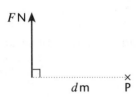
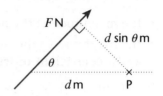

The moment of F about P is
Fd N m ↻.

The moment of F about P is
$Fd \sin \theta$ N m ↻.

When there is more than one force, the **resultant moment** about a point is found by adding the separate moments. One direction is taken to be positive and the other negative.

Example Find the resultant moment about O of the forces shown in the diagram.

Taking ↻ as positive:

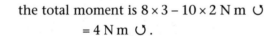

the total moment is $8 \times 3 - 10 \times 2$ N m ↻
$= 4$ N m ↻.

Equilibrium

For an object to be in equilibrium, the resultant force acting on it must be zero and the resultant moment must be zero.

Example

The diagram shows a light rod AB in equilibrium. Find the values of F and d.

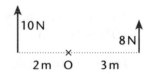

Resolving vertically $\quad F - 30 - 20 = 0$
$\quad\quad\quad\quad\quad\quad\quad \Rightarrow F = 50$

Taking moments about C (written as M(C)) will give an equation involving d.

M(C) gives $\quad 30d - 20 \times 1.5 = 0$
$\quad\quad\quad\quad\quad \Rightarrow d = 1$.

Centre of mass

In the diagram, m_1 and m_2 lie on a straight line through O. Their displacements from O are x_1 and x_2 respectively.

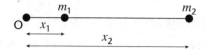

The position given by $\bar{x} = \dfrac{m_1 x_1 + m_2 x_2}{m_1 + m_2}$ is called the **centre of mass** of m_1 and m_2.

In the case where m_1 and m_2 are held together by a light rod, $\bar{x}$ gives the position through which their resultant weight appears to act.
The rod would balance on a support placed in this position.

For n separate masses $m_1, m_2, \cdots m_n$ the position of the centre of mass is given by

$$\bar{x} = \frac{m_1 x_1 + m_2 x_2 + \cdots m_n x_n}{m_1 + m_2 + \cdots m_n}. \text{ This is usually written as } \bar{x} = \frac{\sum_{i=1}^{n} m_i x_i}{\sum_{i=1}^{n} m_i}.$$

Example

In the diagram, the three masses shown lie on a straight line through O. Find the distance of the centre of mass of the system from O.

$$\bar{x} = \frac{3 \times 0 + 2 \times 1 + 5 \times 3}{3 + 2 + 5} \text{ m} = 1.7 \text{ m}.$$

The centre of mass is 1.7 m from O.

In two dimensions, using the usual Cartesian co-ordinates, the position of the centre of mass is at $(\bar{x}, \bar{y})$ where $\bar{x}$ and $\bar{y}$ are given by

$$\bar{x} = \frac{\sum_{i=1}^{n} m_i x_i}{\sum_{i=1}^{n} m_i} \text{ as above, and } \bar{y} = \frac{\sum_{i=1}^{n} m_i y_i}{\sum_{i=1}^{n} m_i}.$$

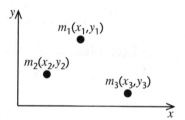

Examiner's Tip

Questions in mechanics frequently involves such things as **light rigid bodies** that simply act as placeholders for individual masses or **uniform** bodies where the mass is taken to be at the geometric centre. A **lamina** is a thin flat object such as a sheet of metal.

Dynamics

Newton's laws of motion

Newton's laws of motion provide us with a clear set of rules that can be used to analyse the effect of forces within a system. The laws may be stated as:

1. Every particle continues in a state of rest or uniform motion, in a straight line, unless acted upon by an external force.
2. The resultant force acting on a particle is equal to its change of momentum. *This law is most often applied in the form $F = ma$.*
3. Every force has an equal and opposite reaction.

The formula $F = ma$

In the formula $F = ma$, F N stands for the resultant force acting on a particle, m kg is the mass of the particle and a m s^{-2} is the acceleration produced.

It is important to use the correct units.

Example

A particle of mass 2 kg rests on a smooth horizontal plane.
Horizontal forces of 15 N and 4 N act on the particle in opposite directions.
Find the acceleration of the particle.

Using $F = ma$
$15 - 4 = 2a \Rightarrow a = 5.5$.

The acceleration of the particle is 5.5 m s^{-2} in the direction of the 15 N force.

The formula applies equally well in 2 or 3 dimensions using vectors.

Example

Forces $(7\mathbf{i} - 2\mathbf{j} + \mathbf{k})$ N and $(3\mathbf{i} + \mathbf{j} - 5\mathbf{k})$ N act on a particle of mass 10 kg.
Find the acceleration produced.

The resultant force is the vector sum of the given forces, so using $\mathbf{F} = m\mathbf{a}$ gives

$(7\mathbf{i} - 2\mathbf{j} + \mathbf{k}) + (3\mathbf{i} + \mathbf{j} - 5\mathbf{k}) = 10\mathbf{a}$
$\Rightarrow \quad 10\mathbf{i} - \mathbf{j} - 4\mathbf{k} = 10\mathbf{a}$
$\Rightarrow \quad \mathbf{a} = \mathbf{i} - 0.1\mathbf{j} - 0.4\mathbf{k}$

The acceleration of the particle is $\mathbf{i} - 0.1\mathbf{j} - 0.4\mathbf{k}$ m s^{-2}.

The formula applies even when the force is variable.

Example

A force $(2t\mathbf{i} - 5\mathbf{j})$ N acts on a particle of mass 0.5 kg at time t seconds. Find the velocity of the particle at time t given that, initially, the velocity is $-3\mathbf{i} + 7\mathbf{j}$ m s^{-1}.

Using $\mathbf{F} = m\mathbf{a}$ gives $2t\mathbf{i} - 5\mathbf{j} = 0.5\mathbf{a} \Rightarrow \mathbf{a} = 4t\mathbf{i} - 10\mathbf{j}$

$$\mathbf{v} = \int 4t\mathbf{i} - 10\mathbf{j}\, dt = 2t^2\mathbf{i} - 10t\mathbf{j} + \mathbf{c}.$$

When $t = 0$, $\qquad \mathbf{v} = -3\mathbf{i} + 7\mathbf{j} \Rightarrow \mathbf{c} = -3\mathbf{i} + 7\mathbf{j}$.

This gives $\qquad \mathbf{v} = 2t^2\mathbf{i} - 10t\mathbf{j} - 3\mathbf{i} + 7\mathbf{j} = (2t^2 - 3)\mathbf{i} + (7 - 10t)\mathbf{j}$.

The velocity of the particle at time t is $(2t^2 - 3)\mathbf{i} + (7 - 10t)\mathbf{j}$ m s^{-1}.

Connected particles

In a typical problem, two particles are connected by a **light inextensible string**. Since the string is light, there is no need to consider its mass.
Since it is inextensible, both particles will have the same speed and accelerate at the same rate *while the string is taut*. The tension in the string acts equally on both particles but in opposite directions.

Example

Two particles P and Q are connected by a light inextensible string. The particles are at rest on a smooth horizontal surface and the string is taut. A force of 10 N is applied to particle Q in the direction PQ. P has mass 2 kg and Q has mass 3 kg. Find the tension in the string and the acceleration of the system.

Using $F = ma$ $\qquad$ For particle P $\qquad\qquad T = 2a$ $\qquad$ [1]
$\qquad\qquad\qquad$ For particle Q $\qquad\qquad 10 - T = 3a$ $\qquad$ [2]

[1] + [2] gives $\qquad\qquad\qquad\qquad\qquad 10 = 5a \Rightarrow a = 2$

Substituting for a in [1] gives $\qquad\qquad T = 4$.

The tension in the string is 4 N and the acceleration of the system is 2 m s^{-2}.

Examiner's Tip

Note the use of the double-headed arrow to represent acceleration.

Dynamics

Momentum

The **momentum** of a particle is a vector quantity given by the product of its mass and its velocity i.e. momentum = mv where m kg is the mass of a particle and v m s^{-1} is its velocity.

The principle of conservation of momentum states that when two particles collide the total momentum before impact = the total momentum after impact.

Before impact
After impact

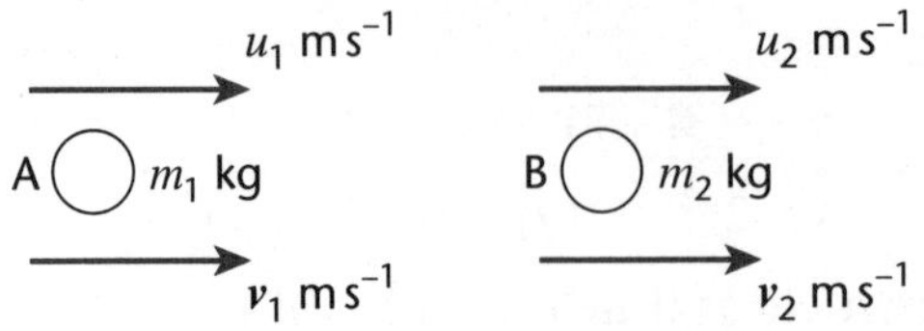

In the diagram, $u_1 > u_2$ so that the particles collide.

Conservation of momentum gives $m_1 u_1 + m_2 u_2 = m_1 v_1 + m_2 v_2$.

Example

A particle of mass 5 kg moving with speed 4 m s^{-1} hits a particle of mass 2 kg moving in the opposite direction with speed 3 m s^{-1}. After the impact the two particles move together with the same speed v m s^{-1}. Find the value of v.

Before impact
After impact

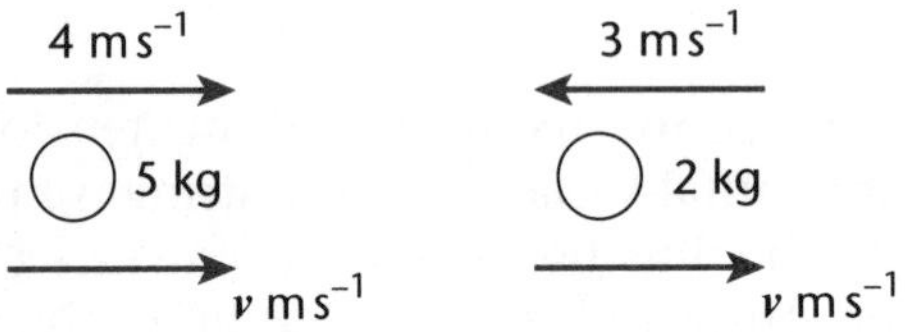

Conservation of momentum gives $5 \times 4 - 2 \times 3 = 5v + 2v \Rightarrow 7v = 14$
$\Rightarrow v = 2$.

Impulse

The **impulse** of a constant force acting over a given time is given by the product of force and time. This may be written as $I = Ft$ where the impulse is I N s, the force is F N and the time is t seconds.

It follows that: impulse = change in momentum

so $Ft = mv - mu$.

Examiner's Tip

When two particles collide, each receives an impulse from the other of equal size but opposite sign. In this way, the total change in momentum is zero as expected.

Example

A particle of mass 4 kg, initially at rest, is acted upon by a force of 3 N for 10 seconds. What is the speed of the particle at the end of this time?

Using $Ft = mv - mu$
gives $3 \times 10 = 4v - 0$
$\Rightarrow \quad v = 7.5.$

The speed of the particle is 7.5 m s^{-1}.

When two particles collide, the contact force between them may only last a very short time and is unlikely to be constant. However, the value of F in the equation $Ft = mv - mu$ may be used to represent the average value of this force.

Example

A ball of mass 0.5 kg strikes a hard floor with speed 2 m s^{-1} and rebounds with speed 1.5 m s^{-1}.
Given that the ball is in contact with the floor for 0.05 s find the average value of the contact force.

Taking upwards as the positive direction:

$u = 2$, $v = -1.5$, $m = 0.5$ and $t = 0.05$.

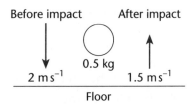

Using $Ft = mv - mu$
gives $0.05F = 0.5 \times 1.5 + 0.5 \times 2$
$\Rightarrow \quad 0.05F = 1.75.$
$\Rightarrow \quad F = 35.$

The average value of the contact force is 35 N.

Examiner's Tip

It is a good idea to draw a diagram and define a positive direction. If you don't do this you may have a problem with minus signs.

Progress check

Vectors

1. Write down the position vector **r** of a point with coordinates (5, −2).
2. Find the magnitude of the vector **r** = 5**i** − 12**j** and give its direction relative to **i**.

Kinematics

1. A stone is thrown vertically upwards with a speed of 15 m s^{-1}. Take the downward acceleration due to gravity to be 10 m s^{-2} and find:
   (a) the greatest height reached
   (b) the speed and direction of the stone after 2 seconds.
2. A particle has position vector **r** = $5t^2$**i** + t^3**j** m at time t seconds. Find its velocity and acceleration when $t = 3$.

Statics

1. An object of mass 8 kg rests on a plane inclined at 40° to the horizontal. Find the components of its weight parallel and perpendicular to the plane. Take $g = 9.81$ m s^{-2}.
2. Find the magnitude and direction of the resultant **R** N of the forces (2**i** + 3**j**) N, (5**i** − 4**j**) N and (−3**i** − 7**j**) N.
3. An object of weight 50 N rests on a rough horizontal surface. A horizontal force of 20 N is applied to the object so that it is on the point of slipping. Find the value of the coefficient of friction.
4. The diagram shows a light rod with masses of 5 g, 2 kg and 1 kg attached. Find the distance of the centre of mass from O.

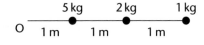

Dynamics

1. A particle of mass 8 kg is acted upon by the forces (11**i** − 3**j**) N, (14**i** + **j**) N and (−**i** + 10**j**) N. Find the acceleration of the particle.
2. A particle of mass 5 kg moves along the x-axis such that its displacement x m from O at time t s is given by $x = t^3 − 15$ for $0 \leqslant t \leqslant 5$.
   (a) Find the acceleration of the particle at time t s.
   (b) Find the resultant force acting on it when $t = 4$.
3. Two particles A and B are connected by a light inextensible string. The particles are at rest on a smooth horizontal surface and the string is taut. A force of 12 N is applied to particle B in the direction AB. A has mass 4 kg and B has mass 2 kg. Find the tension in the string and the acceleration of the system.

Answers on page 95

Statistics 1

Representing data

Measures of central location

An average is a value that is taken to be representative of a data set. The three forms of average that you need are the mean, median and mode. These are often referred to as **measures of central location**.

The **mean** $\bar{x}$ of the values $x_1, x_2, x_3, ..., x_n$ is given by $\bar{x} = \dfrac{\Sigma x_i}{n}$.

If each x_i occurs with frequency f_i then the mean of the **frequency distribution** is given by $\bar{x} = \dfrac{\Sigma f_i x_i}{\Sigma f_i}$.

Example

Find the mean of these results obtained by throwing a dice.

score	1	2	3	4	5	6
frequency	18	17	23	20	24	18

$$\bar{x} = \frac{18 \times 1 + 17 \times 2 + 23 \times 3 + 20 \times 4 + 24 \times 5 + 18 \times 6}{18 + 17 + 23 + 20 + 24 + 18} = \frac{429}{120} = 3.575$$

The **median** is the middle value of the data when it is arranged in order of size. If there are an even number of values then the median is the mean of the middle pair. In the example above, the 60th and 61st values are both 4 so the median is 4.

The **mode** is the value that occurs with the highest frequency. In the example above the mode is 5.

You can *estimate* the mean of **grouped data** by using the mid-point of each class interval to represent the class.

Example

Estimate the mean value of h from the figures given in the table.

An estimate of the mean is given by

$\bar{x} = \dfrac{755}{72} = 10.486...$

$10.5 =$ to 1 d.p.

Interval	frequency (f_i)	mid-point (x_i)	$f_i \times x_i$
$0 < h \leq 5$	8	2.5	20
$5 < h \leq 10$	24	7.5	180
$10 < h \leq 15$	29	12.5	362.5
$15 < h \leq 20$	11	17.5	192.5
Totals	72		755

Examiner's Tip

Once data has been grouped, the exact values are not available and it is only possible to *estimate* the mean.

Mathematics Revision Notes

Representing data

Measures of dispersion

The simplest measure of dispersion, or spread, is the **range** which is the difference between largest value and the smallest value.

A slightly more refined approach is to measure the spread of the 'middle half' of the data. The data is put into order and the **quartiles** Q1, Q2 and Q3 are found that divide the data into quarters.

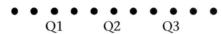

The **inter-quartile range** is then Q3 − Q1.

The quartiles may also be used to indicate whether the data values show **positive skew** (Q2 − Q1 < Q3 − Q2) or **negative skew** (Q2 − Q1 > Q3 − Q2).

$$\text{Variance} = \frac{\sum (x_i - \bar{x})^2}{n} = \frac{\sum x_i^2}{n} - \bar{x}^2.$$

> The variance gives an indication of the spread of data values about the mean but the units of the data have been squared in the process.

$$\text{Standard deviation} = \sqrt{\text{variance}} = \sqrt{\frac{\sum x_i^2}{n} - \bar{x}^2}.$$

> Standard deviation is measured in the same units as the data and is the value most commonly used to measure dispersion at this level.

Statistical diagrams

A box and whisker plot for example shows the location and spread of a distribution at a glance.

A back-to-back stem and leaf diagram allows direct comparison of two sets of data to be made.

The diagram gives a sense of location and spread for each data set.

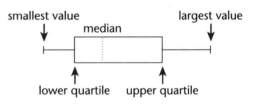

The spread of marks is similar for the boys and the girls but the average for the girls is higher than for the boys.

Examiner's Tip

You can use statistical diagrams to help you compare distributions and reveal further information about the data.

Probability

Venn diagrams

Venn diagrams provide a useful way to represent information about **sets** of objects. The reason for mentioning them here is that the diagrams, and the notation used to express results, have a direct interpretation and application in probability theory.

In each diagram, the rectangle represents the set S of all objects under consideration.

The circles represent particular sets A and B of objects within the set S.

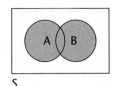

$A \cup B$ represents the set of all objects that belong to *either* A or B or both.

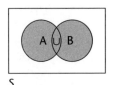

$A \cap B$ represents the set of those objects that belong to *both* A and B.

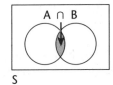

A' represents the set of objects in S that do not belong to A.

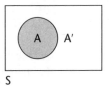

To make use of these ideas in probability theory:

- The objects are interpreted as **outcomes** for a particular situation.
- The set S is the **sample space** of all possible outcomes for the situation.
- The sets A and B are **events** defined by a particular choice of outcomes.
- P(A) for **example**, represents the probability that the event A occurs.
- $P(A \cup B)$ represents the probability that *either* A or B *or both* occurs.
- $P(A \cap B)$ represents the probability that *both* A *and* B occur.
- $P(A')$ represents the probability that A does *not* occur.

The **addition rule** may now be written as:

$$P(A \cup B) = P(A) + P(B) - P(A \cap B).$$

A and B are described as **mutually exclusive events** if they have no outcomes in common. These are events that cannot both occur at the same time.

In this case $P(A \cap B) = 0$ and the addition law becomes

$$P(A \cup B) = P(A) + P(B).$$

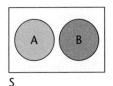

$P(A \cap B) = 0$

Examiner's Tip

Exam questions may be written using a set notation.

Probability

The events A and A' are mutually exclusive for any event A.
Since one of the events A or A' must occur, $P(A \cup A') = 1$ so $P(A) + P(A') = 1$.

This is usually written as $\quad P(A') = 1 - P(A)$.

The **multiplication rule** states that: $P(A \cap B) = P(A | B) \times P(B)$,

$P(A | B)$ means the probability that A occurs *given that* B has *already* occurred.

$P(A | B)$ is described as a **conditional probability**, i.e. it represents the probability of A *conditional* upon B having occurred.

Re-arranging the multiplication rule gives $\quad P(A | B) = \dfrac{P(A \cap B)}{P(B)}$.

If A and B are **independent events** then the probability that either event occurs is not affected by whether the other event has already occurred. In this case, $P(A | B) = P(A)$ and the multiplication rule becomes $P(A \cap B) = P(A) \times P(B)$.

A **tree diagram** is a useful way to represent the probabilities of combined events. Each path through the diagram corresponds to a particular sequence of events and the product rule is used to find its probability. When more than one path satisfies the conditions of a problem, these probabilities are added.

Arrangements and permutations, selections and combinations

There are many situations in probability that can be represented by an **arrangement** of objects (often letters) in a particular order. The following results are useful for calculating the number of possible arrangements.

Description	No of arrangements
n different objects in a straight line.	$n!$
n objects in a straight line, where r objects are the same and the rest are different.	$\dfrac{n!}{r!}$
n objects in a straight line, where r objects of one kind are the same, q objects of another kind are the same and the rest are different.	$\dfrac{n!}{q!r!}$

The number of arrangements that can be made by choosing r objects from n is $^nP_r = \dfrac{n!}{(n-r)!}$. These are known as **permutations**.

The number of **selections** (the order makes no difference) that can be made by choosing r objects from n is $^nC_r = \dfrac{n!}{(n-r)!r!}$. These are known as **combinations**.

Examiner's Tip

Make sure that you are clear about the differences between permutations and combinations.

Correlation and regression

All of the work in this section relates to the treatment of **bivariate data**. This is data in which each data point is defined by two variables.

Correlation

A **scatter diagram** may be used to represent bivariate data. The extent to which the points approximate to a straight line gives an indication of the strength of a linear relationship between the variables, known as the **linear correlation**.

One way to arrive at a numerical measure of the correlation is to use the **product–moment correlation coefficient**, r.

For n pairs of (x, y) values:

$$S_{xx} = \sum x^2 - n\bar{x}^2 \qquad S_{yy} = \sum y^2 - n\bar{y}^2 \qquad S_{xy} = \sum xy - n\bar{x}\bar{y},$$

and the product–moment correlation coefficient is given by: $r = \dfrac{S_{xy}}{\sqrt{S_{xx} S_{yy}}}$.

This gives values of r between -1 (representing a perfect negative correlation) and $+1$ (representing a perfect positive correlation).

An alternative measure of correlation is given by **Spearman's rank correlation coefficient**, r_s. The set of values for each variable must first be ranked from largest to smallest. Some care is needed with equal values:

x	rank
39	1
37	3
37	3
37	3
32	5

The three equal values occupy positions 2, 3 and 4. The average positional value is given by:

$$\frac{2+3+4}{3} = 3.$$

Each of the equal values is given a rank of 3.

At each data point, the x and y values will each have a rank. The difference in these ranks is denoted by d.

The rank correlation coefficient is given by $r_s = 1 - \dfrac{6 \sum d^2}{n^3 - n}$.

This gives values of r_s between -1 (representing a perfect negative correlation) and $+1$ (representing a perfect positive correlation).

Examiner's Tip

The two correlation coefficients given above are comparable but not necessarily equal.

Correlation and regression

Regression

Whereas correlation is determined by the *strength* of a linear relationship between the two variables, **regression** is about the *form* of the relationship given by the equation of a **regression** line.

The purpose in establishing the equation of a regression line is to make predictions about the values of one variable (known as the **response variable**) for some given values of the other variable (known as the **explanatory variable**).

Predictions should only be made for values within the range of readings of the explanatory variable. **Extrapolation** for values outside this range is unreliable. Another factor affecting the accuracy of any predictions is the influence of outliers on the equation of the regression line.

Figure 1 illustrates *y*-**residuals** given by

$d = $ (observed value of y) $-$ (predicted value of y).

Figure 2 illustrates *x*-**residuals** given by

$d = $ (observed value of x) $-$ (predicted value of x).

A **least squares regression line** is a line for which the sum of the squares of either the *x*-residuals or *y*-residuals is minimised.

Figure 1

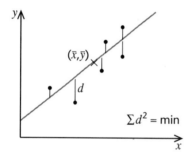

Figure 2

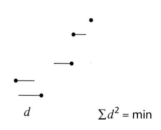

This gives the regression line of y on x as

$$y = a + bx,$$

where $b = \dfrac{S_{xy}}{S_{xx}}$ and $a = \bar{y} - b\bar{x}$.

Use this equation to estimate values of y for given values of x. In this case, x is the explanatory variable and y is the response variable.

This gives the regression line of x on y as

$$x = c + dy,$$

where $d = \dfrac{S_{xy}}{S_{yy}}$ and $c = \bar{x} - d\bar{y}$.

Use this equation to estimate values of x for given values of y. In this case, y is the explanatory variable and x is the response variable.

Unless there is perfect correlation between the variables, the regression lines will be different and you cannot rearrange one equation to obtain the other.

Examiner's Tip

Make sure that you know the difference between correlation and regression.

Discrete random variables

The score obtained when a dice is thrown may be thought of as a **random variable**. Since only specific values may be obtained, this is an example of a **discrete random variable**. In the usual notation:

- A capital letter such as X is used as a label for the random variable.
- A lower case letter such as x is used to represent a particular value of X.
- The probability that X takes the value x is written as $P(X = x)$ or just $p(x)$ and this is known as the **probability function**.
- For a discrete random variable X, with probability function $p(x)$, $\sum p(x) = 1$.

The **probability distribution** of X is given by the set of all possible values of x together with the values of $p(x)$. This is usually shown in a table.

For example, the probability distribution of the scores shown on a dice may be given as:

x:	1	2	3	4	5	6
$p(x)$:	$\frac{1}{6}$	$\frac{1}{6}$	$\frac{1}{6}$	$\frac{1}{6}$	$\frac{1}{6}$	$\frac{1}{6}$

The **cumulative distribution function** is given by $F(x_0) = P(X \leq x_0) = \sum_{x \leq x_0} p(x)$.

This represents the probability that the random variable X takes a value less than or equal to x_0 and is given by the sum of the probabilities up to and including that point.

Expected values

The symbol μ is used to stand for the **mean value of X**. This is also known as the **expected value of X** which is written as $E(X)$.
For any discrete random variable X the mean is $\mu = E(X) = \sum xp(x)$.

Example Find the expected value of the score shown on a dice.

x:	1	2	3	4	5	6
$p(x)$:	$\frac{1}{6}$	$\frac{1}{6}$	$\frac{1}{6}$	$\frac{1}{6}$	$\frac{1}{6}$	$\frac{1}{6}$
$xp(x)$:	$\frac{1}{6}$	$\frac{2}{6}$	$\frac{3}{6}$	$\frac{4}{6}$	$\frac{5}{6}$	$\frac{6}{6}$

$$\mu = \frac{1}{6} + \frac{2}{6} + \frac{3}{6} + \frac{4}{6} + \frac{5}{6} + \frac{6}{6} = \frac{21}{6}$$

The expected value is 3.5.

Examiner's Tip

Re-read this page until you get used to the terms and definitions.

Discrete random variables

The **expected value of a function of a random variable** is found in a similar way. In general, if $f(X)$ is some function of the random variable X then

$$E(f(X)) = \sum f(x)p(x).$$

For example, $E(X^2) = \sum x^2 p(x)$. In the case of the dice scores above this gives:

$$E(X^2) = \tfrac{1}{6} + \tfrac{4}{6} + \tfrac{9}{6} + \tfrac{16}{6} + \tfrac{25}{6} + \tfrac{36}{6} = \tfrac{91}{6}.$$

Using σ to stand for the **standard deviation** of X and $\mathrm{Var}(X)$ to stand for the **variance of** X we have:

$$\sigma^2 = \mathrm{Var}(X) = E(X^2) - \mu^2.$$

Referring to the example of the dice scores again: $\mathrm{Var}(X) = \tfrac{91}{6} - (\tfrac{21}{6})^2 = \tfrac{35}{12}$.

The **expectation of a linear function** of X can be expressed in terms of $E(X)$.
In general: $\quad E(aX + b) = aE(X) + b.$

Once the value of $E(X)$ is known, applying the result above is much simpler than working out $\sum (ax+b)p(x)$. However, the result only holds for linear functions, so for example $E(X^2)$ is not the same as $(E(X))^2$.

The **variance of a linear function** of X may be expressed in terms of $\mathrm{Var}(X)$.
In general: $\quad \mathrm{Var}(aX + b) = a^2 \mathrm{Var}(X).$

The binomial probability distribution

Suppose that n independent trials of an experiment are carried out, each with a fixed probability of success p and corresponding probability of failure q. Then $p + q = 1$ and the probability of r successes is given by the $(r+1)$th term in the binomial expansion of $(q+p)^n$.

Using the random variable X to represent the number of successes in n trials, the probability function is given by $P(X = r) = \binom{n}{r} p^r q^{n-r} \quad r = 0, 1, 2, \ldots, n.$

With this notation, the **binomial probability distribution** may be written as

r:	0	1	2	...	n
$P(X=r)$:	q^n	npq^{n-1}	$\dfrac{n(n-1)}{2} p^2 q^{n-2}$	...	p^n

If X has a binomial distribution with **parameters** n and p then this is written as

$$X \sim B(n, p).$$

In general, if $X \sim B(n, p)$ then $\mu = E(X) = np$ and $\mathrm{Var}(X) = np(1-p)$.

Example

The random variable $X \sim B(8, 0.4)$ find:

(a) $P(X = 2)$ (b) $P(X \leq 2)$ (c) $E(X)$ (d) $Var(X)$.

(a) $P(X = 2) = \binom{8}{2}(0.4)^2(0.6)^6 = 0.2090\ldots = 0.209$ to 3 d.p.

(b) $P(X \leq 2) = P(X = 0) + P(X = 1) + P(X = 2)$
$= (0.6)^8 + 8(0.4)(0.6)^7 + 0.2090\ldots = 0.31537\ldots$
$= 0.315$ to 3 d.p.

(c) $E(X) = np = 8 \times 0.4 = 3.2$.

(d) $Var(X) = np(1 - p) = 8 \times 0.4 \times 0.6 = 1.92$.

The Poisson distribution

The **Poisson distribution** is used to model the number of occurrences of an event in some fixed interval of space or time. It has a single parameter λ which represents the mean number of occurrences in an interval of a particular size. The events occur independently of each other and at random.

If X represents the number of occurrences of the event in an interval of a particular size then:

$$P(X = r) = \frac{e^{-\lambda}\lambda^r}{r!} \qquad r = 0, 1, 2, 3, \ldots \qquad \mu = E(X) = \lambda \qquad \sigma^2 = Var(X) = \lambda.$$

If X has a Poisson distribution with parameter λ then this is written as $X \sim Po(\lambda)$.

Example

The mean number of letters received by a household each day from Monday to Saturday is 3. Find the probability that, on a particular weekday, the number of letters received is (a) 0 (b) at least 2.

If X is the number of letters received in a day then $X \sim Po(3)$.

(a) $P(X = 0) = e^{-3} = 0.0497\ldots = 0.050$ to 3 d.p.

(b) $P(X \geq 2) = 1 - (P(X = 0) + P(X = 1)) = 1 - (e^{-3} + 3e^{-3}) = 1 - 4e^{-3} = 0.8008\ldots$
$= 0.801$ to 3 d.p.

The geometric distribution

The **geometric distribution** is used to model the number of independent trials needed before a particular outcome occurs. Taking X to represent this value and p to be the fixed probability that the outcome occurs:

$$P(X = x) = p(1 - p)^{x-1} \qquad \text{for} \quad x = 1, 2, 3, \ldots.$$

Examiner's Tip

You need to know the conditions under which each model may be applied.

Continuous random variables

A continuous random variable X may take any value in some specified interval. A special function called a **probability density function** (p.d.f.) is needed to describe how the probability is distributed over this interval.

A p.d.f. is usually denoted by $f(x)$ where $f(x) \geq 0$ for all values of x.

Two important properties of a p.d.f. are: $\int_{-\infty}^{\infty} f(x)\,dx = 1$

and $P(a \leq X \leq b) = \int_{a}^{b} f(x)\,dx$.

The Normal distribution

The p.d.f. for the **Normal distribution** is represented by a bell-shaped curve, symmetric about the mean value μ.

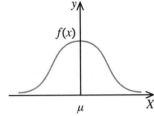

The parameters of the normal distribution are μ and σ^2. If the continuous random variable X follows a Normal distribution with these parameters then this is written as $X \sim N(\mu, \sigma^2)$.

The **standard normal variable** Z is used to calculate probabilities based on the Normal distribution where $Z = \dfrac{X - \mu}{\sigma}$ and $Z \sim N(0, 1)$.

Tabulated values of probabilities for the Normal distribution use Z and the continuous random variable.

Example

Find $P(15 < X < 17)$ where $X \sim N(15, 25)$

$$15 < X < 17 \Rightarrow \frac{15-15}{5} < Z < \frac{17-15}{5} \Rightarrow 0 < Z < 0.4$$

$P(15 < X < 17) = P(0 < Z < 0.4)$
$= \Phi(0.4) - \Phi(0) = 0.6554 - 0.5000$
$= 0.1554$

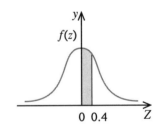

Examiner's Tip

In some questions you may need to use the symmetry of the curve.
For example, it's useful to know that $\Phi(-z) = 1 - \Phi(z)$.

Progress check

Probability

1 The probability that the event B occurs is 0.7. The probability that events A and B both occur is 0.4. What is the probability that A occurs given that B has already occurred?

2 Find the number of arrangements of the letters STATISTICS.

3 Find the number of combinations of seven objects chosen from 10.

Correlation and regression

The summary data for 10 pairs of (x, y) values is as follows:

$$\Sigma x = 146, \quad \Sigma x^2 = 2208, \quad \Sigma y = 147, \quad \Sigma y^2 = 2247, \quad \Sigma xy = 2211.$$

1 Find the value of the product–moment correlation coefficient between x and y.

2 Find the equation of the least squares regression line of y on x.

Discrete random variables

1 The random variable X has the probability distribution shown. Find the value of k.

x:	0	1	2
$p(x)$:	0.3	0.2	k

2 Find (a) $E(X)$ (b) $Var(X)$ for the probability distribution in question 1.

3 The random variable Y is given by $Y = 3X + 2$, where X is as in question 1. Find (a) $E(Y)$ (b) $Var(Y)$.

4 Given that $X \sim B(20, 0.3)$ write down (a) $E(X)$ (b) $Var(X)$.

Continuous random variables

1 A random variable has p.d.f. $f(x)$ given by

$$f(x) = \begin{cases} kx & 1 \leqslant x \leqslant 3 \\ 0 & \text{otherwise} \end{cases}$$

Find the value of k.

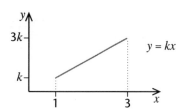

2 Find $P(17 < X < 24)$ where $X \sim N(14, 16)$.

Answers on page 95

Mathematics Revision Notes

Decision mathematics 1

Algorithms

An **algorithm** is a finite sequence of precise instructions used to solve a problem. One way to define an algorithm is to simply list all of the necessary steps. In more complex situations, the process may be easier to follow if the algorithm is presented as a **flow diagram**.

You need to be able to implement an algorithm presented in either form.
You also need to know and be able to implement the following algorithms.

The bubble sort

As its name suggests, the **bubble sort** is an algorithm for sorting a list in a particular order.

Step 1 Compare the first two elements of the list and switch them if they are in the wrong order.

Step 2 Compare the second and third elements of the list and, again, switch them if they are in the wrong order.

Step 3 Continue in the same way until you reach the end of the list. This completes the first **pass** through the list.

Step 4 Make repeated passes through the list until a pass produces no change.

Example   Sort the list 5, 7, 3, 11, 6, 8 into ascending order.

The first pass gives

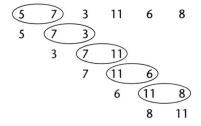

The second pass produces   3, 5, 6, 7, 8, 11.
The third pass produces   3, 5, 6, 7, 8, 11.

The third pass made no change so the ordering is complete.

The quick sort

Step 1 Take the first element in the list as the **pivot**.

Step 2 Consider each of the other elements in turn and place any with order less than or equal to the pivot on one side and any with order greater than the pivot on its other side. Do this without re-ordering the numbers on either side of the pivot.

Step 3 Repeat steps 1 and 2 for each sub-list until each one contains a single element.

Using the quick sort algorithm on the same list as the last example gives:

```
3   5   7   11   6   8
    ↑   6    7   11   8
             ↑    8   11
                  ↑
```

The ordered list is 3, 5, 6, 7, 8, 11.

Binary search

The binary search algorithm provides a systematic way of searching through an ordered list to find an element that matches your criterion. **For example**, you may need to look through a set of index cards to find the contact details of a customer.

Step 1 Find the element in the middle of the list. If this element matches your criterion then stop. If it does not, then use the middle term to decide in which half of the list the required element must lie.

Step 2 Repeat Step 1 for the selected half.

Examiner's Tip

This isn't a very efficient method if the required element is at the end of the list.

Algorithms

Bin packing

The bin packing algorithm is used to solve problems that can be represented by the need to pack some boxes of equal cross-section but different heights into bins, with the same cross-section as the boxes, using as few bins as possible.

There is no known algorithm that will always provide the *best* solution. You need to be familiar with the three algorithms below that attempt to provide a *good* solution. Such algorithms are known as **heuristic** algorithms.

First-fit algorithm

This is the simplest form of bin packing algorithm. It only has one step.
Take each box in turn from the order given and pack it into the first available bin.

Full-bin algorithm

Step 1   Considering the boxes in the given order, use the first available combination that will fill a bin.
Step 2   Repeat Step 1 until no more bins can be filled.
Step 3   Implement the first fit algorithm for the remaining boxes.

First-fit decreasing algorithm

Step 1   Arrange the boxes in decreasing order of size.
Step 2   Implement the first fit algorithm starting with the largest box.

Examiner's Tip

The full-bin algorithm is only practical when the number of bins and boxes is small.

Graphs and networks

- A **graph** is a set of points, called **vertices** or **nodes**, connected by lines called **edges** or **arcs**.
- A **simple graph** is one that has no loops and in which no pair of vertices are connected by more than one edge.

A simple graph

A non-simple graph
Two edges between A and B

A non-simple graph
A loop around C

- The number of edges incident on a vertex is called its **order**, **degree** or **valency**. A vertex may be **odd** or **even** depending on whether its order is odd or even.
- A **subgraph** of some graph G is a graph consisting entirely of vertices and edges that belong to G.
- A **directed** edge is an edge that has an associated direction shown by an arrow. A **digraph** is a graph in which the edges are directed.
- A **path** is a finite sequence of edges such that the end vertex of one edge is the start vertex of the next edge. No edge is included more than once.
- A **connected graph** is one in which every pair of vertices is connected by a path.
- A **complete graph** is one in which every pair of vertices is connected by an edge. A complete graph with n vertices is denoted by K_n.
- A **planar graph** is one that can be drawn in a plane such that no two edges meet except at a vertex.

  K_4 is planar but K_5 is not planar.

- A **cycle** is a path that starts and finishes at the same vertex.
- An **Eulerian** cycle is one that traverses all of the edges of the graph.
- A **Hamiltonian** cycle is one that passes through every vertex once only.
- A **tree** is a graph with no cycles.
- A **spanning tree** is a tree whose vertices are all of the vertices of the graph.
- A **network** is a graph that has a number (**weight**) on every edge.
- A **minimum spanning tree** (MST) of a network is a spanning tree of minimum possible weight. It is sometimes called a **minimum connector**.

Examiner's Tip

It's worth spending time getting to know all of the terms. Their definitions are quite detailed so you will need to keep reminding yourself.

Graphs and networks

Prim's algorithm

Prim's algorithm may be used to find a minimum spanning tree of a network.

Step 1 Choose a starting vertex.
Step 2 Connect it to the vertex that will make an edge of minimum weight.
Step 3 Connect one of the remaining vertices to the tree formed so far, in such a way that the minimum extra weight is added to the tree.
Step 4 Repeat step 3 until all of the vertices are connected.

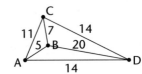

Starting from A gives the MST as

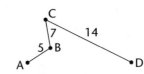

AD could have been used instead of CD

The network given above may be represented by the matrix shown.

A version of Prim's algorithm may be used to find the minimum spanning tree directly from a matrix.

Step 1 Choose a starting vertex. Delete the corresponding row and write 1 above the corresponding column as a label.
Step 2 Circle the smallest undeleted value in the labelled column and delete the row in which it lies.
Step 3 Label the column, corresponding to the vertex of the deleted row, with the next label number.
Step 4 Circle the smallest undeleted value of all the values in the labelled columns and delete the row in which it lies.
Step 5 Repeat steps 3 and 4 until all of the rows are deleted. The circled values then define the edges of the minimum spanning tree.

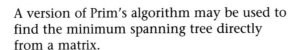

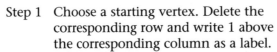

The circled values correspond to the edges, AB, BC, and CD.

Kruskal's algorithm

Kruskal's algorithm is an alternative way to find a minimum spanning tree. In this case, the subgraph produced may not be connected until the final stage.

Step 1 Choose an edge with minimum weight as the first subgraph.
Step 2 Find the next edge of minimum weight that will not complete a cycle when taken with the existing subgraph. Include this edge as part of a new subgraph.

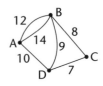

BD can not be included

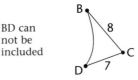

Step 3 Repeat step 2 until the subgraph
makes a spanning tree.

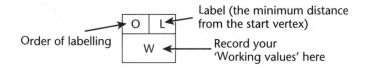

Dijkstra's algorithm

Dijkstra's algorithm is used to find the shortest distance between a chosen *start* vertex and any other vertex in a network.

The algorithm is designed for use by computers but, when implementing it without the aid of a computer, a system of labelling is required.

- O → Order of labelling
- L → Label (the minimum distance from the start vertex)
- W → Record your 'Working values' here

The steps in the algorithm below refer to this system of labelling. Every vertex is labelled in the same way.

Step 1 Give the start vertex a label of 0 and write its order of labelling as 1.

Step 2 For each vertex directly connected to the start vertex, enter the distance from the start vertex as a working value. Enter the smallest working value as a label for that vertex. Record the order in which it was labelled as 2.

Step 3 For each vertex directly connected to the one that was last given a label, add the edge distance onto the label value to obtain a total distance. This distance becomes the working value for the vertex unless a lower one has already been found.

Step 4 Find the vertex with the smallest working value not yet labelled and label it. Record the order in which it was labelled.

Step 5 Repeat steps 3 and 4 until the target vertex is labelled. The value of the label is the minimum distance from the start vertex.

Step 6 To find the path that gives the shortest distance, start at the target vertex and work back towards the start vertex in such a way that an edge is only included if its distance equals the change in the label values.

Examiner's Tip

Reading the letters inside the labelling box as OWL might help you to remember which part is for which information.

Graphs and networks

The travelling salesperson problem

The **travelling salesperson problem** (TSP) is the problem of finding a route of minimum distance that visits every vertex and returns to the start vertex. For a small network it is possible to produce an exhaustive list of all possible routes and choose the one that minimises the total distance. For a large network an exhaustive check is not feasible, even with a computer, because the number of possible routes grows so rapidly.

It is useful to know within what limits the total distance must lie and there are algorithms that can be used for this purpose.

Finding an upper bound

An upper bound for the total distance involved in the travelling salesman problem is given by twice the minimum spanning tree.

The minimum spanning tree (MST) for the network ABCD was found earlier.

The total length of the MST is $5 + 7 + 14 = 26$.

An upper bound for the TSP is $2 \times 26 = 52$. This corresponds to the route ABCDCBA.

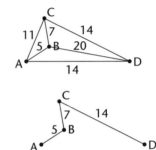

However, an improved (smaller) upper bound can be found by using a **short-cut** from D directly back to A. The route is then ABCDA and the corresponding upper bound is 40.

A different approach is to use the **nearest neighbour algorithm**.

Step 1 Choose a starting vertex.
Step 2 Move from your present position to the nearest vertex not yet visited.
Step 3 Repeat step 2 until every vertex has been visited. Return to the start vertex.

For the network ABCD considered above, the nearest neighbour algorithm gives the same result as using the MST with a short-cut. This will not always be the case.

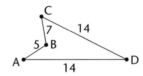

Finding a lower bound

Step 1 Choose a vertex and delete it along with all of the edges connected to it.
Step 2 Find a minimum spanning tree for the remaining part of the network.
Step 3 Add the length of the MST to the lengths of the two shortest deleted edges.

The value found at step 3 is a lower bound for the travelling salesperson problem.

It may be possible to find a better (larger) lower bound by choosing to delete a different vertex initially and repeating the process.

Using the network ABCD again and deleting A gives the minimum spanning tree BCD of length 7 + 14 = 21. Adding the two shortest deleted lengths gives a lower bound of 37.

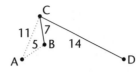

Deleting D to start with gives a lower bound of 40. This corresponds to a cycle ABCDA and so represents the solution of the TSP.

The route inspection problem

The **route inspection problem** is to find a route of minimum total length that traverses every edge of the network, at least once, and returns to the start vertex.

The algorithm for solving this problem is based on the idea of a **traversable** graph. A graph is traversable if it can be drawn in one continuous movement without going over the same edge more than once. If all the vertices of the graph are even, any one of them may be used as the starting point and the same point will be returned to at the end of the movement. Such a graph is said to be **Eulerian**. The only other possibility for a traversable graph is that it has exactly two odd vertices. In this situation, one of the vertices is the start point and the other is the finish point. This type of graph is said to be **Semi-Eulerian**.

The **route inspection algorithm** is:

Step 1 List all of the odd vertices.

Step 2 Form the list into a set of pairs of odd vertices. Find all such sets.

Step 3 Choose a set. For each pair find a path of minimum length that joins them. Find the total length of these paths for the chosen set.

Step 4 Repeat step 3 until all sets have been considered.

Step 5 Choose the set that gives the minimum total. Each pair in the set defines an edge that must be repeated in order to solve the problem.

Examiner's Tip

The algorithm works because repeating the edges between the pairs of odd vertices in this way effectively makes all of the vertices even and creates a traversable graph.

Graphs and networks

Flows in networks

The **flows** referred to may be flows of liquids, gases or any other measurable quantities. The edges may represent such things as pipes, wires or roads that carry the quantities between the points identified as vertices.

A typical vertex has a flow into it and a flow out of it. The exceptions are a **source** vertex which has no input and a **sink** vertex which has no output.

Each edge of the network has a **capacity** which represents the maximum possible flow along that edge. In the usual notation, the capacity is written next to the edge and the flow is shown in a circle.

This edge has a capacity of 10 and a flow of 7.

The set of flows for a network is **feasible** if:

- The total output from all source vertices is equal to the total input for all sink vertices (most will have just one source vertex and one sink vertex).
- The input for each vertex other than a source or sink vertex is equal to its output.
- The flow along each edge is less than or equal to its capacity.

A **cut** divides the vertices into two sets, one set containing the source and the other containing the sink.

The **capacity of a cut** is equal to the sum of the capacities of the edges that cross the cut, *taken in the direction from the source set to the sink set*.
The capacity of cut (i) is $15 + 10 = 25$.
The capacity of cut (ii) is $8 + 14 = 22$.

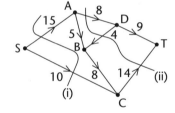

Notice that for the second cut, the capacity of 4 is not included because it does not go from the source set to the sink set.

The maximum flow–minimum cut theorem

The **maximum flow–minimum cut theorem** states that the maximum value of the flow through a network is equal to the capacity of the minimum cut.

It follows from the theorem that if you find a flow that is equal to the capacity of some cut then it must be the maximum flow that can be established through the network.

Once an initial flow has been established through a network it is useful to know the extent to which the flow along any edge may be altered in either direction.

It may be possible to **augment** the flow through the network by making use of this flexibility along a path from the source vertex to the sink vertex. Such a path is called a **flow-augmenting path**.

An algorithm for finding the maximum flow through a network by augmentation is:

Step 1 Find an initial flow by inspection.
Step 2 Label the excess capacity and back capacity for each edge.
Step 3 Search for a flow augmenting path. If one can be found then increase the flow along the path by the maximum amount that remains feasible.
Step 4 Repeat steps 2 and 3 until no flow-augmenting path may be found.

Critical path analysis

The process of representing a complex project by a network and using it to identify the most efficient way to manage its completion is called **critical path analysis**.

Activity networks

A complex project may be divided into a number of smaller parts called **activities**.
The completion of one or more activities is called an **event**.
Activities often rely on the completion of others before they can be started.

The relationship between these activities can be represented in a **precedence table**, sometimes called a **dependency table**.

In the precedence table shown on the right, the figures in brackets represent the **duration** of each activity, i.e. the time required, in hours, for its completion.

A precedence table can be used to produce an **activity network**. In the network, activities are represented by arcs and events are represented by vertices.

The vertices are numbered from 0 at the **start vertex** and finishing at the **terminal vertex**.

Activity	Depends on
A(3)	–
B(5)	–
C(2)	A
D(3)	A
E(3)	B, D
F(5)	C, E
G(1)	C
H(2)	F, G

The direction of the arrows shows the order in which the activities must be completed.

There must only be *one* activity between each pair of events in the network. The notation (i, j) is used to represent the activity between events i and j.

A **dummy activity** is one that has zero duration. A dummy is needed in this network to show that G depends on C whereas F depends on C *and* E.

A dummy is shown with a dotted line. Its direction is important in defining dependency. In this case is shows that F depends on C *not* tht G depends on E.

The **earliest event time** for vertex i is denoted by e_i and represents the earliest time of arrival at event i with all dependent activities completed. These times are calculated using a **forward scan** from the start vertex to the terminal vertex.

The **latest event time** for vertex i is denoted by l_i and represents the latest time that event i may be left without extending the time for the project. These times are calculated using a **backward scan** from the terminal vertex back to the start vertex.

Critical path analysis

The **critical path** is the longest path through the network. The activities on this path are the **critical activities**. If any critical activity is delayed then this will increase the time needed to complete the project. The events on the critical path are the **critical events** and for each of these $e_i = l_i$.

It is useful to add the information about earliest and latest times to the network. The critical path is then easily identified.

The **total float of an activity** is the maximum time that the activity may be delayed without affecting the length of the critical path. It is given by:

latest finish time – earliest start time – duration of the activity.

Scheduling

The process of allocating activities to workers for completion, within all of the constraints of the project, is known as **scheduling**.

The information regarding earliest and latest times for each activity is crucial when constructing a schedule. This information may be presented as a table or as a chart.

Activity	Duration	Start Earliest	Start Latest	Finish Earliest	Finish Latest	Float
A(0, 1)	3	0	0	3	3	0
B(0, 2)	5	0	1	5	6	1
C(1, 3)	2	3	7	5	9	4
D(1, 2)	3	3	3	6	6	0
E(2, 4)	3	6	6	9	9	0
F(4, 5)	5	9	9	14	14	0
G(3, 5)	1	5	13	6	14	8
H(5, 6)	2	14	14	16	16	0

The critical activities are shown along one line.

The diagram illustrates the degree of flexibility in starting activities B, C and G. Remember that G cannot be started until C has been completed.

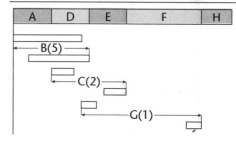

Examiner's Tip

The purpose of scheduling is often to find the number of workers needed to complete the project in a given time, or the minimum time required for a given number of workers to complete the project.

Linear programming

Formulating a linear programming problem

To formulate a linear programming problem you need to:

- Identify the **variables** in the problem and give each one a label.
- Express the **constraints** of the problem in terms of the variables. You need to include non-negativity constraints such as $x \geq 0$, $y \geq 0$.
- Express the quantity to be optimised in terms of the variables. The expression produced is called the **objective function**.

Example

A small company produces two types of armchair. The cost of labour and materials for the two types is shown in the table.

	Labour	Materials
Standard	£30	£25
Deluxe	£40	£50

The total spent on labour must not be more than £1150 and the total spent on materials must not be more than £1250. The profit on a standard chair is £70 and the profit on a deluxe chair is £100.
How many chairs of each type should be made to maximise the profit?

In this case, the variables are the number of chairs of each type that may be produced. Using x to represent the number of standard chairs and y to represent the number of deluxe chairs, the constraints may be written as:

$$30x + 40y \leq 1150 \Rightarrow 3x + 4y \leq 115$$
$$25x + 50y \leq 1250 \Rightarrow x + 2y \leq 50$$

and $x \geq 0$, $y \geq 0$.

Using P to stand for the profit, the problem is to maximise $P = 70x + 100y$.

The graphical method of solution

Each constraint is represented by a region on the graph.

The blue line represents the points where the profit takes a particular value. Moving the line in the direction of the arrow corresponds to increasing the profit. This suggests that the maximum profit occurs at the point X.

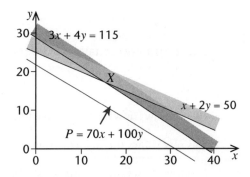

Examiner's Tip

It is a good idea to shade out the *unwanted* region for each one. The part that remains unshaded then defines the **feasible region**.

Linear programming

Solving $3x + 4y = 115$ and $x + 2y = 50$ simultaneously gives X as $(15, 17.5)$.

The nearest points with integer co-ordinates in the feasible region are $(15,17)$ and $(14,18)$. The profit, given by $P = 70x + 100y$, is greater at $(14,18)$.

The maximum profit is made by producing 14 standard and 18 deluxe chairs.

The Simplex algorithm

When there are 3 or more variables, a different approach is needed. The **Simplex algorithm** may be used to solve the problem algebraically. The information must be expressed in the right form before the algorithm can be used.

- First write the constraints, other than the non-negativity conditions, in the form $ax + by + cz \leq d$.
- Write the objective function in a form which is to be maximised.
- Add **slack variables** to convert the inequalities into equations.
- Write the information in a table called the **initial tableau**.

Example

Find the maximum value of $P = 2x + 3y + 4z$ subject to the constraints:

$$3x + 2y + z \leq 10 \quad \text{[i]}$$
$$2x + 5y + 3z \leq 15 \quad \text{[ii]}$$
$$x \geq 0, y \geq 0, z \geq 0.$$

Using slack variables s and t, the inequalities [i] and [ii] become:

$$3x + 2y + z + s = 10$$
$$2x + 5y + 3z + t = 15.$$

The objective function must be rearranged so that all of the terms are on one side of the equation to give:

$$P - 2x - 3y - 4z = 0.$$

The information may now be put into the initial tableau.

	P	x	y	z	s	t	value
The top row shows the objective function	1	-2	-3	-4	0	0	0
This row shows the first constraint	0	3	2	1	1	0	10
This row shows the second constraint	0	2	5	3	0	1	15

The columns for P, s and t contain zeros in every row apart from one. In each case, the remaining row contains 1 and the value of the variable is given in the end column of that row. The value of every other variable is taken to be zero. So, at this stage, the tableau shows that $P = 0$, $s = 10$, $t = 15$ and x, y and z are all zero. This corresponds to the situation at the origin.

Using the algorithm is equivalent to visiting each vertex of the feasible region until an optimum solution is found. This occurs when there are no negative values in the objective row.

When the present tableau is not optimal, a new tableau is formed as follows:

- The column containing the most negative value in the objective row becomes the **pivotal column**. In this case the pivotal column corresponds to the variable z.
- Now divide each entry in the *value* column by the corresponding entry in the pivotal column provided that the pivotal column entry is positive.
  For the example above this gives $\frac{10}{1} = 10$ and $\frac{15}{3} = 5$
  The smallest of these results relates to the bottom row which is now taken to be the **pivotal row**. The entry lying in both the pivotal column and the pivotal row then becomes the pivot. In this case, the pivot is 3.

P	x	y	z	s	t	value
1	−2	−3	−4	0	0	0
0	3	2	1	1	0	10
0	2	5	3	0	1	15

pivotal row

- Divide every value in the pivotal row by the pivot, this makes the value of the pivot 1. The convention is to use fraction notation rather than decimals.
- Then turn the other values in the pivotal column into zeros by adding or subtracting multiples of the pivotal row.

P	x	y	z	s	t	value
1	$\frac{2}{3}$	$\frac{11}{3}$	0	0	$\frac{4}{3}$	20
0	$\frac{7}{3}$	$\frac{1}{3}$	0	1	$-\frac{1}{3}$	5
0	$\frac{2}{3}$	$\frac{5}{3}$	1	0	$\frac{1}{3}$	5

pivotal row

- If there are no negative values in the objective row then the **optimum tableau** has been found. Otherwise choose a new pivotal column and repeat the process. Each time through the process is called an **iteration**.

This tableau shows that the maximum value of P is 20 and that this occurs when $x = 0$, $y = 0$, $z = 5$, $s = 5$ and $t = 0$.

Examiner's Tip

In this case the optimum tableau has been found after just one iteration. It is often necessary to repeat the process in order to find the optimum tableau.

Matchings

Matchings and graphs

A **bipartite graph** is a graph in which the vertices are divided into two sets such that no pair of vertices in the same set is connected by an edge.

In this case, the two sets are {A, B, C} and {p, q, r, s}.

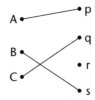

Some vertices in a bipartite graph may not be connected to another vertex.

A **matching** between two sets may be represented by a bipartite graph in which there is at most one edge connecting a pair of vertices.

A **maximal matching** is a matching which has the maximum number of edges. This occurs when every vertex in one of the sets is connected to a vertex in the other set. The bipartite graph shown above represents a maximal matching.

A **complete matching** is a matching in which every vertex is connected to another vertex. This can only occur when the two sets contain the same number of vertices.

The matching improvement algorithm

Figure 1 is a bipartite graph showing the possible connections between two sets. It does not represent a matching because some vertices have more than one connection.

Figure 2 is a bipartite graph representing an initial matching.

An initial matching may be improved by increasing the number of connections. This is the purpose of the **matching improvement algorithm**.

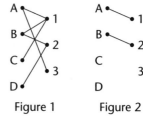

Figure 1   Figure 2

- **Step 1** Start from a vertex not connected in the initial matching and look for an **alternating path** to a vertex in the other set that is not connected.
- **Step 2** Each edge on the alternating path, not included in the initial matching, is now included and each edge originally included is removed.
- **Step 3** Repeat steps 1 and 2 until no further alternating paths can be found.

Alternating path   Maximal matching

Examiner's Tip

In an alternating path, the edges *alternate* between those that are not in the initial matching and those that are.

Progress check

Algorithms

1. Write the letters P, Q, B, C, A, D in alphabetical order using:
   (a) the bubble sort algorithm
   (b) the quick sort algorithm.

2. A project involves activities A–H with durations in hours as given in the table.

A	B	C	D	E	F	G	H
3	1	5	4	2	3	4	2

The project is to be completed in 8 hours.
(a) Use the first-fit algorithm to try to find the minimum number of workers needed.
(b) Find a better solution.

You can represent each worker as a 'bin' or 'height' 8 hours.

Graphs and networks

(a) Use Prim's algorithm to find the total weight of a minimum spanning tree for this network.

(b) Verify your answer to part (a) using Kruskal's algorithm.

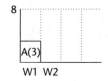

Critical path analysis

Determine the critical activities and the length of the critical path for this network.

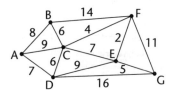

Linear programming

An initial simplex tableau for a linear programming problem is shown

P	x	y	z	s	t	value
1	−3	−2	−1	0	0	0
0	2	3	1	1	0	10
0	3	4	2	0	1	18

(a) Carry out one iteration to produce the next tableau.
(b) Explain why the tableau found is optimal.
(c) Write down the maximum value of P and the corresponding values of the other variables.

Answers on page 95

Progress check answers

Pure 1

Algebra

1. (a) 6, 0
   (b) $f(x) = (x-1)(x+4)(3x+2)$
   (c) $x = 1, -4$ or $-\frac{2}{3}$
   (d)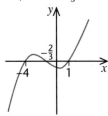

2. (a) $3 \pm \sqrt{2}$ (b) 0.17, −3.84
3. $A = 3, B = 5, C = -2$

Sequences and series

1. (a) 2, 8, 18, 32, 50
   (b) 10, 32, 98, 296, 890
2. 500500
3. 4115

Trigonometry

1. (a) One-way stretch with scale factor 0.5 parallel to the x-axis.
   (b) One-way stretch with scale factor 3 parallel to the y-axis.
   (c) One-way stretch with scale factor 0.5 parallel to the x-axis, followed by a one-way stretch with scale factor 3 parallel to the y-axis, followed by a translation of 1 unit downwards.

2. $\sin^2 x + 2\sin x \cos x + \cos^2 x$
   $+ \sin^2 x - 2\sin x \cos x + \cos^2 x$.
   $\equiv 2(\sin^2 x + \cos^2 x) \equiv 2$

3. 200.6°, 20.6°, −159.4°, −339.4°

Co-ordinate geometry

(a) $3x + 2y = 23$
(b) $y - 6 = \frac{2}{3}(x + 2)$ or $3y - 2x - 22 = 0$.

Differentiation

1. Tangent: $y = 2x + 18$;
   normal: $y - 36 = -\frac{1}{2}(x - 9)$.
2. Local maximum $(-1, 15)$;
   local minimum $(3, -49)$.

Integration

1. (a) 2.5 (b) $12\frac{2}{3}$ (c) 72
2. $y = x^4 + x + 7$

Pure 2

Algebra and functions

1. (a) $fg(x) = 4x^2 + 12x + 14$
   (b) $fg(10) = 534$
   (c) $x = -0.08452, -5.915$.
2. $1 + 36x + 594x^2$
3. $x = 3.322$

Trigonometry

1. RHS $\equiv \dfrac{2\tan\theta}{\sec^2\theta} \equiv 2\tan\theta \times \cos^2\theta$
   $\equiv 2\sin\theta\cos\theta \equiv \sin 2\theta \equiv$ LHS.
2. $x = 30°, 150°$ or $270°$
3. (a) $13\sin(x + 1.176)$
   (b) $x = -0.607$ or $x = 1.397$ to 3 d.p.

Differentiation

1. (a) $10e^x$ (b) $4e^x - 3x^2$ (c) $8e^{2x}$
2. (a) $\dfrac{1}{x}$ (b) $\dfrac{1}{x} - 3e^x$
   (c) $-11\sqrt{2}(1 - \sqrt{2}x)^{10}$

Integration

1. (a) $3\ln|x| + c$ (b) $\ln|x| - \dfrac{3}{x} + \chi$
   (c) $3e^x - \ln|x| + c$
2. $\displaystyle\int_1^3 \pi e^x \, dx = \pi[e^x]_1^3 = \pi(e^3 - e)$.

Numerical methods

1. (b) 2.09 (c) 2.0867
2. 2.31

Mechanics 1

Vectors
1. $r = 5i - 2j$
2. $|r| = 13$, $67.4°$ clockwise.

Kinematics
1. (a) 33.75 m
   (b) 5 m s^{-1} downwards
2. $v = 30i + 27j$, $a = 10i + 18j$

Statics
1. parallel: 50.4 N, perp: 60.1 N
2. 8.94 N at $-63.4°$ to i
3. 0.4
4. 1.5 m.

Dynamics
1. $3i + j$ m s^{-2}.
2. (a) $6t$ m s^{-2} (b) 120 N.
3. Tension = 8 n, acceleration = 2 m s^{-2}.

Statistics 1

Probability
1. 4/7
2. $10! \div (3! \times 3! \times 2!) = 50\,400$
3. 120

Correlation and regression
1. 0.799
2. $y = 2.317 + 0.8481x$

Discrete random variables
1. $k = 0.5$
2. (a) 1.2 (b) 0.76
3. (a) 5.6 (b) 6.84
4. (a) 6 (b) 4.2

Continuous random variables
1. The area under the graph is given by $\dfrac{3-1}{2}(k + 3k) = 4k$
   Since $f(x)$ is a p.d.f.
   $4k = 1 \Rightarrow k = 0.25$
2. 0.2204

Decision mathematics 1

Algorithms
1. (a) A, B, C, D, P, Q
3. (a) First fit: 4 workers
   (b) A better solution is
       (A, B, D) (C, F) (E, G, H)

Graphs and networks
(a) 30

Critical path analysis
A, D, G, J
Length = 21

Linear programming
(a)

P	x	y	z	y	t	value
1	0	$\frac{5}{2}$	$\frac{1}{2}$	$\frac{3}{2}$	0	15
0	1	$\frac{3}{2}$	$\frac{1}{2}$	$\frac{1}{2}$	0	5
0	0	$-\frac{1}{2}$	$\frac{1}{2}$	$-\frac{3}{2}$	1	3

(b) There are no negative values in the objective row.
(c) $P = 15$, $x = 5$, $y = 0$, $z = 0$, $s = 0$, $t = 3$.

Mathematics Revision Notes

Index

acceleration 50, 53–56, 62–63
activity network 87
addition rule 69
algebra 15–15, 30–37
algorithms 78–80, 82–85, 90, 92
angle formulae 39–40
area under a curve 28, 48
arithmetic progression 16–17
average 67
binary search 79
binomial expansion 33–35
binomial probability distribution 74
bipartite graph 92
bounds 84–85
box and whisker plot 68
bubble sort 78
calculus 24–28
capacity 86
centre of mass 61
chain rule 43
collision 64–65
combinations 70
component form vectors 51–52, 57
compound angle formulae 39–40
conditional probability 70
connected particles 63
conservation of momentum 64
constant acceleration 53, 55
continuous random variable 76
co-ordinate geometry 21–23
correlation 71
correlation coefficient 71
cosecant 38, 40
cosine 18–19, 38–41
cotangent 38
critical path 87–88
cubic curve 10
cumulative distribution 73
curve sketching 25
cut 86
data 67–69
decimal search 46
dependency table 87
derivative 24–26
differentiation 24–26, 27, 42–43
Dijkstra's algorithm 83
direction 50, 52, 55
discrete random variables 73–75
dispersion 68
displacement 50, 53–54
distance 50, 54
domain 6, 30, 38
double angle formulae 40
earliest event time 87
equilibrium 58–60
Eulerian 81, 85
expected value 73–74
exponential equations 37
extrapolation 72
factor theorem 14, 36
factorising 7, 14–15
feasible region 89
first-fit algorithm 80
flow augmentation 86
flow diagram 78
flows 86
force 50, 57–58, 62–65
force diagram 57
frequency distribution 67
friction 57–59
friction coefficient 59
full-bin algorithm 80
function 5–6, 30–33
geometric distribution 75

geometric progression 16–17
geometry 21–23
gradient 21–23, 24, 36
graphical terms 81
gravity 53, 56, 57
grouped data 67
half angle formulae 40
identity 13–14, 39–41
image 5
impulse 64–65
inclined plane 57
independent events 70
indices 5
inequalities 15
integral 28
integration 27–28, 37, 44–45, 48
integration constant 27
intercept 21–23
inter-quartile range 68
inverse of a function 30–32
inverse trigonometric functions 39
iteration 47, 91
kinematics 53–56
Kruskal's algorithm 82
lamina 61
latest event time 88
least squares regression 72
light inextensible string 63
light rigid bodies 61
log laws 37, 42
logarithm 37, 42
magnitude 50, 52, 58
map 5
mass 50, 57, 59, 61–65
matching improvement algorithm 92
maximum 25–26
maximum flow–minimum cut theorem 86
mean 67, 73, 76
median 67
mid-point 23
minimum 25–26
mode 67
modulus function 11, 32–33
moments 60
momentum 50, 62, 64
multiplication rule 70
mutually exclusive events 69–70
natural logarithm 37, 42
nearest neighbour algorithm 84
Newton's laws of motion 62–63
normal 25
Normal distribution 76
normal reaction 57
parameters 75
Pascal's triangle 34–35
periodic function 18, 36
permutations 70
pivot 79
pivotal column and row 90–91
point of inflexion 25–26
Poisson distribution 75
polynomials 14, 35
position vector 51, 55
precedence table 87
Prim's algorithm 82
probability 69–70, 73–75
probability density function 76
probability distribution 74
programming 89–91
projectiles 56
Pythagoras' theorem 18
Pythagorean identities 38–41

quadratic graphs 9–10
quadratics 6–8, 13–15, 41
quick sort 79
quotient 35
radians 20
random variable 73–76
range 6, 30, 70
reaction 57
reciprocal function 44
recurrence relation 16
recursive definition 16
reflection 32
regression 72
remainder 35–36
remainder theorem 35–36
residuals 72
resolving a vector 51–52
resolving forces 57, 59–60
resultant force 58, 62
resultant vector 50
route inspection problem 85
scalar multiplication 51
scalar quantity 50
scatter diagram 71
scheduling 88
secant 38, 40
second derivative 26
sequences 16
series 16–17
sets 69
Simplex algorithm 90–91
simultaneous equations 13, 22–23, 63
sine 18–19, 38–41
skew 68
speed 50, 54–55
square root 5
standard deviation 68, 74, 76
standard normal variable 76
stationary points 25–26
statistical diagrams 95
statistics 67–77
stem and leaf diagram 68
straight line equation 21–23
straight line motion 53
substitution 13, 21–22
surds 5
symmetry 10, 18, 38
tableau 90–91
tangent 18–19, 38–40
tangent to a curve 25
tension 57, 63
time 50, 53–54, 64–65
total float 88
transformations 12, 19
trapezium rule 48
travelling salesperson problem 84–85
tree diagram 70
trigonometry 18–20, 38–41
turning effect 60
uniform bodies 61
unit vectors 51
variable acceleration 54–55
variance 68, 74, 76
vector diagrams 57
vectors 50–52, 55
velocity 50, 53–56, 63–65
venn diagrams 69
vertex 9, 81–85, 87, 91, 92
volume of revolution 45
weight 57–58